Inventology

How We Dream Up Things That Change the World

Pagan Kennedy

BANTAM PRESS

LONDON • TORONTO • SYDNEY • AUCKLAND • JOHANNESBURG

TRANSWORLD PUBLISHERS
61–63 Uxbridge Road, London W5 5SA
www.transworldbooks.co.uk

Transworld is part of the Penguin Random House group of companies
whose addresses can be found at global.penguinrandomhouse.com

First published in the US in 2016
by Houghton Mifflin Harcourt Publishing Company
3 Park Avenue, 19th Floor, New York, New York 10016

First published in Great Britain in 2016 by Bantam Press
an imprint of Transworld Publishers

Book design by Mark Robinson

A CIP catalogue record for this book
is available from the British Library.

ISBNs 9780593073476 (tpb)
9780593077238

Typeset in Bell MT
Printed and bound by Clays Ltd, Bungay, Suffolk.

1 3 5 7 9 10 8 6 4 2

To my grandfather Stephen Patrick Burke, inventor

CONTENTS

Introduction ix

Part I: Problem Finding

CHAPTER 1: MARTIAN JET LAG 3

CHAPTER 2: USER-INVENTORS 12

CHAPTER 3: SOMEONE ELSE'S SHOES 23

CHAPTER 4: THE FUTURE OF FEEDBACK 31

Part II: Discovery

CHAPTER 5: SUPER-ENCOUNTERERS 47

CHAPTER 6: DATA GOGGLES 63

CHAPTER 7: BUILDING AN EMPIRE OUT OF NOTHING 79

Part III: Prophecy

CHAPTER 8: THE PONG EFFECT 97

CHAPTER 9: THE WAYNE GRETZKY GAME 108

CHAPTER 10: THE MIND'S R&D LAB 125

CHAPTER 11: HOW TO TIME-TRAVEL 137

Part IV: Connecting

CHAPTER 12: THE GO-BETWEENS 157

CHAPTER 13: ZONES OF PERMISSION 168

CHAPTER 14: HOLISTIC INVENTION 180

Part V: Empowerment

CHAPTER 15: PAPER EYES 193

CHAPTER 16: TINKERING WITH EDUCATION 206

Conclusion 216

Acknowledgments 225

Notes 227

Index 256

INTRODUCTION

In 2012, the *New York Times Magazine* hired me to write the weekly "Who Made That?" column, and so I began hunting down the people behind such inventions as sliced bread, the 3-D printer, and lipstick. Week after week, I discovered that ideas pop up wherever they please, and inventors come from every corner of life. A test pilot created aviator sunglasses; a frustrated father devised the sippy cup to foil his own toddler; and experiments in a kitchen in Queens, New York, led to the Xerox machine.

Why did these people hit on solutions that everyone else had missed? That question became a thread that I followed to an even bigger one: Is there a formula for invention, and if so, can anyone learn it?

So when I reached out to creative people for my column, I asked them about the process that had led them to conceive their bold ideas. One of the first people I queried, Jake Stap, taught me that sometimes we discover the power of our imagination only when we feel desperate. In the late 1960s, Stap worked as a coach at two Wisconsin tennis camps, and during those long days of lessons, he spent hours stooped over, retrieving hundreds of balls. His back ached, and he urgently needed to invent his way out of this drudgery.

Stap placed a tennis ball on the passenger seat of his car, where it rolled around for weeks, a reminder that he needed to think about the problem. As he drove, he performed mental experiments: he built an imaginary tennis court in his mind's eye and pictured what it would be like to wear an arm extender that would let him reach the ground without bending over. But, he realized, the mechanical hand would pick up only one ball at a time. That was no good. Finally, during one of his meditations, Stap reached over and pinched the tennis ball on the seat next to him. When the rubber yielded under his fingertips, he had a new idea: the ball could squeeze through metal bars, taking a one-way trip into a wire bin.

Stap rigged up a basket with a handle and metal bars across its bottom so he could perform real-life experiments. "I fooled around with the bars to find the right distance" so that the balls would pop into the basket and stay there, he told me. He called his invention a "ball hopper."

The next summer at tennis camp "everyone wanted to use them," recalled his daughter, Sue Kust. "It was a mad dash for the ball hoppers." She observed that "when people saw how [the ball hopper] worked and how simple it was, they would always say, 'I could have thought of that.'"

His concept might have *seemed* obvious, and yet it wasn't. Rubber balls became standard tennis equipment in the 1870s, so by all rights, a Victorian gentleman should have used a wire basket to wrangle his tennis balls. Instead, for nearly a century, tennis players chased down the rubber balls without ever seeing Stap's solution. That's the mystery that surrounds some inventions. They seem easy in hindsight. And yet, the most elegantly simple breakthroughs can hide from us for decades. So what blocks us from grasping an idea? And how can we find the "obvious" ideas that are hiding from us right now?

To answer those questions, we must study many inventions and look for patterns. For instance, if you delve into the history of cancer cures, squirt guns, and smoke detectors, you'll find startling similarities in the way that they came into existence. So if we can observe the techniques that have led lots of inventors to success, we might extrapolate what methods work best.

Invention versus Innovation

People tend to use the words *invention* and *innovation* interchangeably, which causes confusion. And so before we proceed any further, we should come up with working definitions for both. Art Fry—the originator of the Post-it Note—developed his own way of distinguishing invention from innovation, and his definitions are so illuminating that I will borrow them and use them throughout this book. Invention, according to Fry, is what happens when you translate a thought into a thing. More specifically, Fry points out that an invention usually involves creating a prototype that lets you test your concept and demonstrate that it works. Once you've created that model, "the creation becomes an invention," according to Fry. The process may require dreaming, drawing, observation, idea generation, discovery, tinkering, and engineering. But it should end with the proof.

Innovation is what happens afterward. It "is the act of working through all of the obstacles and problems in the path of turning a creative idea into a business," according to Fry. Indeed, the term *innovation* is often used as a catchall word to describe the challenges companies must overcome in order to mass-produce a product—like streamlining, shaving costs, managing supply chains, and assembling teams of collaborators. The business side of product development is an art unto itself. But this book, for the most part, will not concern itself with business innovation.

Instead, we will investigate the first steps, the embryos, origins, and private visions that give birth to new things. As I spoke with inventors, they told me stories about hunting the original thought as if it were a rare bird flitting through the forest. The process often involves a craftsmanship of the imagination, in which we carry out experiments in our fantasies. "When I get an idea I start at once building it up in my imagination. I change the construction, make improvements and operate the device in my mind," the visionary inventor Nikola Tesla wrote. He was describing a process of mental iteration that we all possess—but that too few of us have truly learned how to use.

Why We Need "Inventology"

This book will focus on what might be called "micro-creativity," that is, invention at the level of the individual. I'm not trying to advance what is known as the Great Man Theory, in which lone heroes are given sole credit for a breakthrough. Rather, it is an acknowledgment that you are one person, and so am I. While it's interesting to find out which city generates the most patents per capita (Eindhoven, in the Netherlands, is often at the top of the list), that doesn't give us much insight into what we need to do, as individuals, to become more imaginative; after all, if you buy a plane ticket to Eindhoven and stroll along its charming canal, you likely won't be struck with genius.

It's crucial that we find out what people actually *do* as they invent things. What are they doing in their minds and with their hands? We need a new field of study—call it Inventology—to answer that question. If you aspire to run a marathon, you can read thousands of books on training for peak performance; you can pore over studies of the benefits of carb loading and wind sprints. But for those who aspire to invent, it's much harder to find this kind of actionable research.

And yet, as I dug through historical archives, I encountered a few pioneers who did try to discover a formula for invention. For instance, a Soviet science-fiction writer named Genrich Altshuller pored over thousands of patents in the mid-twentieth century, mining that trove for clues about the human imagination. He developed methods for predicting future technologies and for solving mechanical riddles. He also founded a school for inventors in Azerbaijan the likes of which has never existed before or since. Later in this book, we will spend time with Altshuller and some of the other visionaries who tried to launch a new science of invention.

We will also meet the modern-day researchers whose work can help us understand the inventive mind. They're economists, psychologists, inventors, neuroscientists, engineers, crowdfunders, and ethnographers. Because these investigators are isolated from each other in different fields, the puzzle pieces of Inventology are widely scattered. This book will put those pieces together. It is based on more than a hundred interviews with inventors and explorers in many fields, as well as dozens of studies and research papers.

I have set out to answer four questions:

1. Who *really* invents?
2. How do they do it?
3. What can the rest of us learn from the data on successful invention?
4. How will twenty-first-century invention be transformed by crowdfunding, 3-D printing, big data, and other new technologies?

A bit more about that final question. We are at a moment in history when the barriers to invention are falling as never before. On your laptop, you can draw upon R&D tools far beyond anything that a Bell Labs engineer would have had in his workshop in the 1960s. You can raise money from a bunch of strangers, and then ask them to give you feedback. You can exchange digital files that encode the shape of an eyeglass lens or the curve of a bike frame. You can communicate directly with factories and operate like a commercial manufacturer. You can use your phone and a credit card to hire lab researchers who will test a drug on a genetically engineered mouse according to your directions. You can consult a worldwide library of millions of research papers and exchange ideas with zillions of potential collaborators.

Many of the people I interviewed for this book alerted me to how profoundly their lives had changed because of these new tools. Their personal experiences speak to the revolutionary shift now taking place.

In the 1870s, Thomas Edison built an idea factory where he corralled engineers, mechanics, and chemists and leaned over their shoulders. That centralized, invention-in-one-place method caught fire in the twentieth century, but now it seems to be going the way of the incandescent light bulb. Already many of us are becoming, at least in some small degree, inventors. We can act as small-time funders of products. We can tell corporations about what we want and even collaborate with them on design, entering into a two-way communication about the things that we use. If we hate a product, we can collectively kill it on a site like Amazon, where we band together with other people who point out its shoddy design. And we can form communities to invent everything from sports equipment to body parts.

The tools are changing, and so too is the kind of imagination that we will need to master the new opportunities around us. Ideas are no longer just hovering in the air; they're also zipping through fiber-optic cable. For this reason, I will focus on inventions and discoveries of the past fifty years, rather than those from earlier eras.

Five Paths

We tend to believe that great ideas arrive like angels, in a flash of light. This assumption has been handed down to us from the ancient Greeks, who regarded creativity as a gift from the Muses; as the ancients saw it, people don't invent so much as wait for a deity to deliver an illumination. In the Middle Ages, the word *inspiration* meant that God had breathed the truth straight into a person's mind. Even today, we talk of problem solving as a passive process of gobsmacking—and we treasure stories about revelations that came easily and took only a moment.

One of those fables concerns August Kekulé, the chemist who dreamed about a snake eating its own tail and woke up to discover the ring-shaped structure of the benzene molecule. But that story, so often repeated as fact, was popularized by a humorous article written in the nineteenth century, and Kekulé's famous dream was likely the punch line to a joke about scientific posturing. Still, even when we know these stories are false, they enchant us.

Perhaps that's because the true narratives of invention are nuanced and labyrinthine, and they don't lend themselves to fairy tales. The wizards I interviewed, like Bill English (who worked with Doug Engelbart on the first computer mouse), were eager to instruct me about just how much time it takes to crack a difficult problem—you have to observe, imagine, fantasize, forage, and experiment just to cobble together the *idea*, never mind the prototype.

They described several different paths toward the breakthrough, some of which had never occurred to me. For instance, an engineer named Martin Cooper, inventor of the hand-held cell phone, told me that his first inklings began in the 1960s with a vision of a science-fiction future. Cooper imagined that one day everyone would be issued a phone number at birth and walk around with communicators in their pockets.

In the 1970s, as smaller batteries and transistors became available, Cooper and his colleagues at Motorola did manage to cobble together a hand-held cell phone. Primitive though it was, the prototype opened up a new set of possibilities when Cooper staged a theatrical stunt on the sidewalk of Sixth Avenue in Manhattan; he paced around, yelling into the gizmo, and nearly collided with a cab, drawing a crowd of amazed New Yorkers who had never before witnessed this kind of behavior. But even after Cooper proved the technology could work, it would take Motorola another decade to commercialize the first hand-held cell phone.

What Cooper described to me was the opposite of a eureka situation. He began with a vision of the impossible, and then he deployed his imagination like a movie director or novelist to time-travel into the future. Indeed, many technologies start out like the plot of a science-fiction story. And so that is just one of the routes that inventors take to prove that their "impossible" idea is actually inevitable.

I have divided this book into five parts to reflect various strategies that inventors deploy on the way to success. Each part tells a story about one kind of imagination and how it can be used to overcome challenges and root out hidden opportunities.

Part I delves into *problem finding*. We will look at how creative people—like Jake Stap—use their frustration as a doorway into the imagination. According to an old saw, necessity is the mother of invention; that's certainly true, but the adage is annoyingly vague. What kind of necessity works best to help reveal the outlines of a hidden problem? Why do some frustrations lead to a big idea, while most don't? And can we learn from someone else's pain?

Of course, not every invention begins when someone recognizes a new problem. Some inventors work backward; they stumble across a surprise—a sound, a flavor, or a clue in the data—and realize that it could be the answer to a well-known problem. In Part II, we will look at *discovery* and consider the role of serendipity in the creative process. In 1928, Alexander Fleming returned to his lab after a vacation and found mold growing in some of his Petri dishes; he might have washed away the mess, but instead he peered at the dishes under a microscope. In 1929, Fleming published a paper describing the antibiotic action of the mold, and that inspired others to develop penicillin as a medication. So how do accidents turn into inventions? And is it possible to use new tools—like big data—to supercharge the rate of serendipity?

In Part III, we will examine the strategy of *prophecy* and *futuristic thinking*. In one of his novels, Jules Verne sent explorers to the moon in a bullet-shaped capsule, and by doing so, he encouraged millions of other people to dream of exploring space. Through science fiction, we can jolt ourselves into finding new possibilities. This is particularly true in the fields of computing and communications, where the spin of technological improvement is so fast that the future arrives in a matter of months. How can we think ahead of the curve? Are there laws that govern the way that technology evolves? What kind of imagination is required to predict the future?

In Part IV, we'll delve into the challenge of *connecting* unusual ideas together. We will meet the people who act as cross-pollinators, buzzing from one domain to another, carrying ideas with them like pollen. Here, we will investigate the mental skills required to bring together two seemingly incompatible ideas. Who are the matchmakers? And how do they unite a problem and a solution that otherwise would not emerge? We'll also find out how new tools are bringing together unlikely collaborators and ensuring that the best ideas rise to the top.

In Part V, we will explore the challenges of *empowerment*. It takes enormous courage to claim a problem. When you dare to tackle a big question, you may face ridicule, rejection, and opposition. So how do you grant yourself permission to invent? And how do educators teach children to challenge the status quo and take possession of the designed environment? In this final part of the book, we will contemplate the future of the imagination itself, and the political and social implications of a world in which billions of people have access to sophisticated R&D tools.

PART I

PROBLEM FINDING

1

MARTIAN JET LAG

IN 1970, BERNARD D. SADOW, A VICE PRESIDENT AT A LUGGAGE company, was schlepping two suitcases through an airport when he noticed a workman pushing a machine on a dolly. Inspired, he began to experiment with a rolling suitcase that looked like a large pull toy; eventually he patented a suitcase that sat squarely on rollers, with a flexible strap attached to it. Instead of carrying this suitcase, you pulled it behind you on a "leash." Sadow's idea was revolutionary—here was one of the first suitcases designed for airports.

Though it sold well in the 1970s, his suitcase didn't end up becoming standard equipment for the air traveler. You rarely see pull-toy luggage today. Why not? Sadow's design was only a half solution. When you pulled too hard, the suitcase would crash into your legs. If you yanked it around a corner, it might lose its balance and flop onto its side.

In the 1980s, a pilot named Robert Plath custom-built his own version of the rolling suitcase in his home workshop. His design was a vast improvement over Sadow's. Plath put wheels on one edge of the bag so that it could tip on its corner, and he outfitted it with a rigid handle. You could adjust the length of the handle by sliding it up or down, trombone-style, allowing you to find just the right angle so the bag would follow you obediently, without attacking your ankles. This was a bag that you could comfortably tote over miles of airport linoleum.

So why was a pilot's insight so much more fruitful than the executive's? The answer has something to do with the way the two men experienced the problem. Bernard Sadow, a businessman heading off on vacation, was merely a tourist looking for a better solution. His was a short-term form of necessity. But Plath—who dragged his bags to and fro after every shift, day after day—was motivated to think deeply about the suitcase problem, to tinker in his garage, and to come up with an ingenious design for frequent flyers. By virtue of his job, Plath was already living in the future, when flying would become a commonplace misery.

In the 1990s, the price of airline tickets plummeted. Companies began sending executives around the world, sometimes on three or four flights a week. Planes began to feel like buses—crowded, smelly, and raucous. "Life Sucks and Then You Fly," as one *Wired* headline put it, in an article that described tech employees suffering in the middle seats during their coast-to-coast commute. By that time, passengers were hunting for anything that would ease the pain of cramped flights—from Xanax to noise-canceling headphones. And that's when the rolling suitcase became essential equipment. Plath's Rollaboard suitcase took off.

Adam Smith, writing in *An Inquiry into the Nature and Causes of the Wealth of Nations* (1776), observed that there is a special kind of magic in tasks that we repeat over and over again. He described a pin factory where one man straightened the wire, another man cut it, and yet another man sharpened the tip, and so on. In a factory like that, each laborer became an expert in one small task, and his close attention might inspire him to "find out easier and readier methods of performing" his job.

In fact, Smith argued that one of the side benefits of the factory system was the way it turned workmen into inventors. He praised the "pretty machines" that factory laborers devised to ease their drudgery. For instance, he noticed a boy who was supposed to pump a lever in time with a piston. This relentless, grinding task inspired the boy to figure out an ingenious work-around: he tied a string between the lever and a moving part elsewhere on the machine. Now the machine itself pulled the lever for him. After automating his job, the boy skipped off to play with friends.

The economist Eric von Hippel, speaking in 2005, made his own observation about the way repetition can feed the imagination: "I've learned personally that you can get a graduate student to do a lot of things, but you can't get them to do it twenty thousand times in a row, [because] they will start to invent" a way to automate the boring job. There seems to be some kind of threshold—some number of hours—after which frustration produces creative insight.

In the 1970s, von Hippel came up with a name for the people who struggle with problems for which no off-the-shelf solution is available: he dubbed them Lead Users. Their job or hobby exposes them to an unusual kind of repetition, tedium, or danger. When bike hobbyists began to spend hours out in the woods riding over boulders and tree stumps, their tires popped, and that inspired them to build what we now call mountain bikes. Surgeons who pioneered new methods of operating on the heart had to design tools in order to perform these feats. And in 1982, a professor at Carnegie Mellon University recognized a new problem with digital communication—the flame war—so he devised the happy-face symbol, or emoticon, to cool tempers online.

Lead User Theory

Before he joined academia and became a professor at MIT's Sloan School of Management, von Hippel worked as an engineer at a start-up. And that's how he discovered the existence of Lead Users. In the 1960s, he became one himself.

Back then, von Hippel needed a tiny fan that would allow him to improve the performance of a fax machine, so he contacted an aerodynamics expert at Princeton and together they designed the fan. With his plans in hand, von Hippel struck a deal with a manufacturing company to produce the device.

Soon, von Hippel received a call from someone at the manufacturer: "It turns out a lot of other people want your fan too," the company rep told von Hippel. "Can we . . . produce it for them?"

Von Hippel said yes. And then one day he picked up an industry journal and noticed an advertisement for his fan. The company had claimed credit for inventing it. You'd think he might have been angry. But instead, he was fascinated. He had just stumbled across a clue—the first inkling of an insight that would change his life, as well as what we know about technological creativity.

In the 1970s, when he switched careers and became an academic researcher, he dedicated himself to a question: Who *really* dreams up breakthrough ideas? To find out, he came up with a method that bears a startling resemblance to the way that a detective works a cold-case murder—digging deep into files, interviewing witnesses, and wearing down shoe leather to follow clues. In one of his earliest studies, von Hippel picked more than a hundred lab equipment products and then hired researchers to help him discover the backstory of each of the devices. He learned that about 80 percent of the scientific equipment products had begun with someone who needed the tool. For instance, at a Harvard conference in 1964, a lab worker described a method he'd invented to "bake away" the dirt on a microscope using a piece of gold foil; later that year, a manufacturer transformed this concept into a product. Subsequent studies—by von Hippel and others—have shown that the pattern holds true in many other fields.

A company may have "begged, borrowed, stole[n], or bought [its] idea from a person who never becomes famous," Dr. Nat Sims, the inventor-in-residence at Massachusetts General Hospital, told me. Companies then "invest a few hundred million dollars into making [the product] successful and getting over all the hurdles. So it becomes an integral part of their culture—not for any mean or malicious reason—to forget that history" of the product. After a few years pass, no one knows how the product came to be—and the true origin story is very hard to uncover. Eventually, we all believe that the product started with the manufacturer.

Of course, only certain kinds of problems are valuable. Ideally, you would want to suffer from a frustration that is rare now (so that no one else knows about it) but that will one day bother lots of people. "Lead Users are familiar with conditions which lie in the future for most others," von Hippel wrote, and so they "can serve as a need-forecasting laboratory." And some Lead Users experience a problem so futuristic that the rest of us have trouble even imagining it. Take Martian jet lag, a sleep disorder that bedevils the engineers who work with Mars-based equipment.

Because the Martian day is slightly longer than ours, people who control robots on the Red Planet have to continually shift their schedules, eating breakfast at 3:40 a.m., then 4:20 a.m., then 5:00 a.m. "It feels like you are perpetually flying east 40 minutes every day," said one scientist, Deborah Bass. "It starts to take its toll."

To add to the discomfort, each Mars landing mission operates in its own time zone to correspond to the local sunrises and sunsets on the planet. For this reason, Scott Maxwell, an engineer and driver on the Mars Exploration Rover mission, had to consult a spreadsheet and then perform several calculations to figure out when he needed to wake up. In 2012, Maxwell created the MarsClock phone app to help him track the Mars rovers and get to work on time. Writing the app, he told me, "scratched two itches: it gave me a handy Mars-time alarm clock, and it let me share a bit of the fun of the mission with rover fans, something I'm always seeking ways to do." Thousands of fans did download the app—an artifact from the Mars mission that lived on their phones.

Engineers like Maxwell face all kinds of other problems that don't affect anyone else right now—like what to do when dust clogs a machine located on another planet. And many of their inventions are one-off solutions that will never spread widely. But imagine what would happen if a company decided to install a vast robot-run mine on Mars; at that point, thousands of us might be complaining about our damned droid with its burned-out wires. And if that future does come, who will have pioneered the solutions to the problem of interplanetary robot control? Most likely, it will be the scientists who first grappled with the problem. In Part III of this book, we will dig deeper into the question of futuristic problems and see why prediction and forecasting are crucial to the inventing process.

But for now, let's return to the subject at hand—problem finding—and recap what we know so far. The most valuable kind of frustration has three components:

1. It plays out over a long period of time, thus inspiring more and better solutions.
2. It reveals a hidden problem that is difficult to detect.
3. It forecasts a problem that will affect thousands or millions of people in the future.

As it happens, all three kinds of frustration came together in Jack Dorsey.

Pre-Tweet

As a kid in the 1980s, Dorsey loved to listen to the chatter on CB radio and police scanners, and he became fascinated with the way that drivers of fire trucks and ambulances had developed their own lingo;

the drivers spoke in short coded blasts, telling each other where they were located and what they were doing.

By 2000, Dorsey had found a job as a code jockey, writing dispatch software to help route cars and trucks around city streets. He was still passionate about traffic, so much so that he developed a strange desire. If an ambulance could announce its whereabouts and activities, why couldn't he? Dorsey began to imagine a kind of dispatch software for himself, one that would work something like a police scanner, bleating out his activities as he moved around San Francisco and Oakland.

"I wasn't considering what everyone else wanted. I was considering what *I* wanted," Dorsey later told a reporter. So he cobbled together some software just to satisfy this private desire. At the time, Dorsey owned the RIM 850, one of the first phones that could display and send e-mail messages, and he devised a method to send out text-based broadcasts.

One day, in Golden Gate Park, he sent out a dispatch intended to alert friends to his location and what he was doing (watching the bison). The message was met with silence. Few of his friends owned the type of phone that would allow them to receive the broadcast Dorsey had sent out. This was his moment of Martian jet lag; he experienced a problem years ahead of everyone else. And yet, Dorsey felt confident that other people would catch up.

Six years later, now at a company called Odeo, Dorsey explained his concept to his coworkers. By then, the world *had* caught up with Dorsey, and millions of phones were enabled with the Short Message Service (SMS) protocol, which made it easy to send and receive texts.

Dorsey and his collaborators hacked together their social network in two weeks. In the beginning, Twitter was something like Bernard Sadow's suitcase. Sadow had recognized the potential of transforming luggage into a vehicle on wheels, but his execution of the concept had been awkward. Likewise, Dorsey and his collaborators at Twitter had given entirely new capabilities to the cell phone—turning it into a twenty-first-century CB radio on which impromptu communities could spring up to report on unfolding events. And yet, the site was awkward to use in its first incarnation; it crashed frequently, and it lacked many of the features that now make it so addictive.

The Bucket Brigade

Earlier in this chapter, we saw that pain and frustration have a cumulative effect on invention—as Adam Smith noted, when people begin repeating the same operations over and over, they learn an enormous amount about how to remove the drudgery and unpleasantness from machines.

The story of Twitter—and many other social platforms—suggests that the hours of frustration do not have to be experienced by a single person. If thousands of people repeat the same unnecessary keystroke, even if they spend only a few seconds a day doing it, one of those users will notice this nano-frustration and will invent a way around it.

So while Twitter's engineers struggled to put out their fires and keep the site running, a bucket brigade of users were improving the functionality of the site and tailoring it to fit their needs. For instance, in 2006, a user named Robert Andersen added an @ sign to the front of his brother Buzz's name to indicate that he was addressing a remark directly to Buzz. Other users embraced the idea, because it was so useful. "If you look at all the early tweets, there are no conversations until people started using the @ symbol," Dorsey himself acknowledged.

As users spent thousands—then millions—of hours on the site, they accumulated lots of insights about its failings and their own frustrations. Those hours of drudgery or exasperation had enormous value. The Twitter citizens began to experiment with their own lingo and thumb-saving shortcuts—not just the @ sign, but also the hashtag and the retweet. This suggests a fourth principle connected to necessity and invention: if you want to understand a problem, ask a community.

When lots of people experience the same kind of pain or frustration, they generate an enormous amount of information about unsolved problems. But if that knowledge is scattered across thousands of different people's minds, how do we gather it together again?

In the chapters that follow, we will investigate how communities of people come together to articulate their own needs and define problems. We will also look at how inventors can work with those communities to extract the most valuable insights from them.

2

USER-INVENTORS

IN THE 1990S, TIM DERK SPENT A LOT OF TIME INSIDE A FUR suit and a foam head, prancing around basketball arenas as the Coyote mascot for the San Antonio Spurs. That was back in what might be called the "slingshot era," a time when team mascots used enormous rubber bands to fling souvenirs into the crowd during games. The slingshots had limited range, and Derk was frustrated because he couldn't reach the fans at the top of the arena. So he worked for months to find a better way.

"It weighed ninety pounds, including the tanks," said Derk of the T-shirt cannon he debuted for fans in the 1990s. (Derk is hazy on the exact date.) "It was like carrying a TV set on your back. The gun was probably at least four feet long. It used a cast-iron pipe—the kind that goes into the floor underneath your commode," he told me. Fans adored it, and the T-shirt gun went viral, spreading to arenas around the country.

Derk didn't care whether he received credit for his "invention"—as he saw it, no one in particular owned the idea. "It wasn't that one minute it did not exist, and then it did. It just evolved" out of the mascot community, Derk told me. "The Phoenix Gorilla and I were two of the pioneers," but many of their colleagues also contributed to the stunt. Professional mascots are, as he puts it, a "fur-ternity." Everyone shares gimmicks; they borrow, riff, and vie to come up with the wackiest spin on any particular idea.

I'm not sure whether Kenn Solomon considers himself to be a member of the fur-ternity, but he also believes in sharing and improvising. Solomon, the mascot for the Denver Nuggets, wears an enormous yellow head to perform as Rocky the Mountain Lion. He too is constantly inventing. "If your hand is a fluffy paw, you can't run a camera," he told me. "You can't pick up a pencil. You don't have any grip to throw a football or a basketball." He solved the paw problem by rebuilding his own costume so that he can extend and retract his (human) fingers through small holes in the claws. Now, he signs autographs and can even "pick up babies and hold newborns without being afraid of them slipping."

Mascots are always trying to solve a problem: How do you project a personality from inside thirty pounds of plush and foam? And that pushes them to think hard about their equipment. "I would never build a mascot costume where you don't look out of the eyes," Solomon said with passion in his voice. "Mascots look awkward with the mouth open all the time and having to hold their head back so they can see. You can't develop a character that way. You've got to look out of the eyes."

Tim Derk and Kenn Solomon possess knowledge about human movement and costume making that few other people can match. But they don't presume to own any of this expertise. They operate in a free space where sharing is the norm. Patent lawyers have a term for it: the "negative spaces" are the zones in which people create without seeking to patent their ideas. Agriculture, folk songs, magic tricks, religions, jokes, hairstyles, Wikipedia, languages, and roller-derby *noms de guerre*—many of the great human endeavors have emerged from the negative spaces. In this common realm, we come together to share our worries and find solutions together, as members of a community.

The Internet itself is a negative space, because it belongs to no one and is continuously revamped by everyone. It affords an entirely new ecosystem—a virtual universe—that nourishes the weeds and wildflowers spawned by millions of minds; it's like a vast R&D lab where we all share our own experiments and benefit from the work of others. For instance, a friend of mine, who was constructing an elaborate dollhouse for his daughter, discovered hundreds of instructional videos on YouTube dedicated to this one task. After starting his project, he fell down a rabbit hole into the world of doll enthusiasts who have come up with ingenious ways to build tiny mansions.

The Dark Matter of Invention

The negative space is so enormous, and involves so many millions of people, that it's hard to measure, and even harder to figure out how much value it generates. Eric von Hippel calls it "the dark matter," because the creative ferment is all around us and yet difficult to quantify. He and his colleagues have pointed out that this inventive activity in the garages and basements—which is then shared on the Internet—probably dwarfs the efforts that companies put into R&D to develop their products.

Organizations like the National Science Foundation track spending on R&D by universities, companies, nonprofits, and state agencies, but there has been little effort to measure the output of makers, hobbyists, and open-source inventors. Hundreds of public hackerspaces have popped up in the past decade, yet it's difficult to find a comprehensive list of them all.

However, we do know a bit about the *people* who tinker because demographers have begun to map out the invisible nation of gearheads. A 2010 survey of British citizens found that 8 percent of the population modify and prototype their own tools and technologies. According to a 2013 *Time* magazine survey of thousands of adults in seventeen countries, *one-third* of respondents called themselves inventors, suggesting that billions of people take pride in the way they have improved, built, or tweaked the designed environment. This could include creating or modifying products at home, from Ikea furniture hacks to phone apps to go-carts. The dark matter out there appears to be enormous, and it contains a potential mother lode of information about the needs and problems that have not been addressed by off-the-shelf products.

Of course, some people tinker just for the sake of tinkering. But many others do so because they haven't been able to find a product that satisfies their needs. Dr. Nat Sims, the inventor based at Massachusetts General Hospital, told me that some of his doctor friends maintain their own machine shops; when they notice a problem with a surgical tool, they will often reinvent it or design a new one on their own. They do this because they need equipment that simply doesn't exist, and it's easier to make it themselves than to convince a company to manufacture it for them.

Of course, there are many inventions that simply cannot come from home workshops. Some (which we will look at later on in this book) must be developed by institutions with long timelines and deep pockets. The cell phone, the transistor, the laser—these are the kinds of breakthroughs that are discovered in the lab and that cost millions or billions to research and develop. But in the area of applied inventions—simple fixes that satisfy a need—the users often have an advantage over companies and academic labs. They're the ones who experience the long-term frustration that reveals opportunities for invention. The home inventors step in to fill niches that the professional designers have neglected.

This is spectacularly evident in the world of prosthetics, where a community of volunteers is rising up to create affordable devices for themselves and those in need.

Debra Latour was born with an arm that ends just below the elbow, so when she was a toddler, doctors fitted her with a mechanical arm with pincers in place of a hand.

Back then, in the early 1960s, prosthetic devices were primitive and clunky. As a girl, Latour had to strap herself into a harness that wound around one shoulder and a cable that ran down her arm to control the pincers.

When she was little, Latour believed some inventor would eventually free her from the awkward harness; surely by the time she'd grown up she would be wearing a space-age device. And indeed, engineers at the Defense Advanced Research Projects Agency (DARPA) and other cutting-edge labs *have* developed bionic arms that mimic the look and feel of flesh; but these devices can be staggeringly expensive—upward of $100,000. Our health insurance system often doesn't help patients to pay for the high-end devices, according to Latour. And she points out that simple mechanical equipment tends to be more reliable than the high-tech gear.

That's why in 2005, she was still lashing herself into a harness every morning. "When I wanted to grasp an object, I had to pull forward on my opposite shoulder in order to create tension on the cable that ran down my arm," she told me. Decades of performing that motion had damaged her body and caused chronic pain.

I asked her why the mechanical prosthetics hadn't improved over those decades. Why didn't someone come up with a better solution?

"The thing about people with upper-limb deficiencies is that we are a niche group," Latour explained. "We're a very small population." With so little profit to be made, companies weren't offering a lot of options.

For Latour, the turning point came when she attended a spiritual retreat. While meditating there, she realized that she couldn't wait around for someone else to solve her problem. She began imagining ways to improve the harness, and then one night, about two weeks after the retreat, "I woke up with an uneasy feeling," she said. A vivid image took shape in her mind; in her imagination, she was examining her own back, the pale skin glimmering with portent. As she concentrated on this image, a pushbutton materialized underneath her right shoulder blade. "I realized, 'Oh, my goodness! I could replace the entire harness with a small plastic patch.'" The next day, she began experimenting with cables and plastic parts. Within three months, she had engineered a device the size of a drink coaster that she called the Anchor. She affixed it next to the skin near her shoulder blade. By flexing a small muscle in her back—and using that muscle almost like the tip of a finger—she could push against the catch on the Anchor; that triggered a cable, which in turn would control the grip of her mechanical hand. She no longer needed to wear a harness. And she was no longer in pain.

Latour offered the device as an option to patients at the Shriners Hospital for Children in Springfield, Massachusetts, where she then worked as a clinician; she discovered that it worked well for dozens of people with upper-body differences. She donated the intellectual property to Shriners Hospital, and then the hospital granted Latour permission to manufacture and market the Anchor. Now Latour and her husband run a business out of their dining room, assembling the devices and sending them out to people who need them. "We don't do it for the money," Latour told me; instead, she is driven by a passion to help other people escape from their harnesses.

When I first met Latour in the summer of 2014, she had driven for hours through a rainstorm to Brown University to speak at Superhero Cyborgs workshop, a weeklong program for kids interested in engineering. Dressed in a crisp white shirt, black pants, and sneakers, with her blond hair pulled back into a ponytail, Latour projected the capable air of someone who works with patients. That day, she was wearing her favorite arm—a carbon-fiber tube that ends in a pair of strong pincers. The kids gathered around as she demonstrated the precision of her gripping mechanism. She put one foot up on a chair, untied her shoelace, and retied it; her movements were fast and graceful.

"Wow," marveled a teenage boy in the group. "That hand might not look human but it really gets the job done."

"Exactly," Latour said. "And it's not just the hand that's important. You've got to power it with your body."

Here Latour turned around to show the kids the bump beneath her shirt—this was the Anchor device that she'd been using to control the grip of her pincers. "What I'm trying to tell you is that sometimes the simplest ideas are the very best."

She was teaching them a lesson in survival. The cost of many upper-body prosthetic devices (like carbon-fiber arms) is still so high that many patients can't afford them—they can run from $5,000 to $80,000. Parents are faced with heartbreaking dilemmas: Would you spend, say, $7,000 for a hand for your daughter even though she'll grow out of it in a year or two?

Even in the United States, kids with upper-body differences often have to make do without mechanical hands. That dilemma has inspired a community of children, parents, and craftspeople to come together. There are now thousands of people around the world designing and sharing body parts.

In 2012, a professor at the Rochester Institute of Technology, Jon Schull, attended a conference where engineering students talked about how they had designed and built a custom prosthetic device for a patient. "Everyone else felt great about it," Schull told me—but he began to worry. He remembers asking a question at that meeting: "If there's one guy who needs that prosthetic device, there are probably another hundred thousand people around the world who have the same problem. Are we doing anything to help them?"

His colleagues told him, "Actually, no one is dealing with that problem."

So Schull began to imagine a global network of universities that would provide 3-D-printed prosthetic devices to anyone who needed them. "While I was sitting in that session it just became clear to me that that should exist," he told me. "I tried to get colleagues at other universities to collaborate on this. But it didn't work. It's hard to get people at different schools to work together. I couldn't even get my own university excited about it. So after a few months I gave up."

A year later, Schull saw a YouTube video about a South African carpenter named Richard Van As; after an accident in which he sliced off two of his fingers, the carpenter shopped for prosthetic fingers and hands—and discovered that the devices were so expensive that he couldn't afford them. That inspired him to reach out to Ivan Owen, a puppeteer and gadget maker based in Washington State; together, the two men created the Robohand, a mechanical hand that could be assembled from parts manufactured on a 3-D printer. The fingers of the hand had nylon cord "tendons" that tightened when the wearer bent his or her wrist; this allowed the hand wearer to grip, for instance, a pencil or a glass of water. The two men made the digital design plans for the Robohand free to all, so that anyone in the world with access to a 3-D printer could manufacture it for under $30.

When Schull found out about the Robohand, he realized that any studio outfitted with a 3-D printer could now become a prosthetics shop. All it took was for someone to organize volunteers and pair them with those in need. "I had what was, in retrospect, a very clever, very naive idea," Schull said. He made a Google map showing where volunteers with 3-D printers could offer their help; meanwhile, parents who desired a custom-made Robohand for a child could also put a pin in the map. And that's how the e-NABLE community formed.

Since then, the community has mushroomed to more than thirty-two hundred volunteers; as I write this, e-NABLE volunteers have made more than seven hundred prosthetic devices, serving children (and some adults) who wouldn't otherwise have access to mechanical hands.

Schull dreamed that the community could provide an affordable solution to patients in need, but he didn't foresee the enormous diversity and creativity that would emerge from this process. In the e-NABLE community, kids are often designers, and since the parts are 3-D-printed in plastic, the child can choose any color—green, purple, pink. The young client can have as many fingers as he or she wants, even specifying a preference for embossed stars or leather accents. The hand can be as smooth as a pony hoof or as scaly as a dragon's claw.

"Greg Dennison—he owns his own 3-D printer, and he invented a really cool hand for his son. It has a thumb on each side" to give the boy a stronger grip, Schull told me. Another child, named Tully, wanted a glow-in-the-dark arm—and got it. "And I had a kid sitting on my lap three weeks ago, a nine-year-old named Derek from Buffalo. I was showing him this arm contraption that we were developing for him. But as I'm trying to explain it to him, he picks up two of our models, puts them together, and constructs a double-length arm. He said, 'I want my arm to be *that* long.' And why not?"

The e-NABLE community gives us a glimpse into what inventing can look like when manufacturing and collaboration become ubiquitous. At that point, it becomes possible to satisfy even the most quirky individual desire. "This group has unleashed a Cambrian explosion of new hand designs," Schull said. "There are no managers and no factories and no central plan." The most impressive breakthrough is the price: "The commercial market sometimes charges forty-freaking-thousand dollars for a mechanical hand. They rarely cost less than a few thousand dollars. And we're making them for free."

Until recently, if you suffered from a problem, you either had to buy a solution or build it yourself. But now a third way is opening up. "There is now a real possibility of distributed, decentralized invention, manufacturing, and distribution," Schull prophesied. "The tools of production are now in the hands of the masses. There's a real chance that we're going from the Industrial Revolution to the Information Revolution to the Alternative Economy Revolution."

And yet, we still talk about invention as if it emerges only from inside the Edison-style labs under the direction of corporate planners. It's an assumption so ingrained, so much a part of the way we think, that it shows up in the way we speak: "*They* should make a milk carton that doesn't leak," we say. *They* are engineers and designers whose job is to understand us. We call regular people "consumers," while companies are "producers."

But many of us now want to interact with products in the same way we engage the media—that is, to exist in the gray area between consumer and producer. Economists sometimes use a portmanteau word to describe this new species: *prosumers.* The *Oxford Dictionaries* give an intriguing definition of this word as "a prospective consumer who is involved in the design, manufacture, or development of a product or service." Frankly, I'm not thrilled with the term *prosumer*, because it suggests that our creativity is just another facet of our lives as consumers. Maybe the best word continues to be *inventor*—if we could use that word in a wider and more inclusive way.

As we have seen in this chapter, communities of people who share the same pains and desires can come together to discover and define problems, and the Internet is making this kind of collaboration far more practical. Of course, it's not always possible to connect with the users—or benefit from their wealth of knowledge—online. That's particularly true of the disadvantaged and underserved communities that may not have access to computers. And even when people are wired up, it can be difficult to truly understand their needs unless you spend time getting to know them. This is why inventors often immerse themselves in other people's troubles and pains.

Some engineers and designers even embrace the practice of ethnography, spending weeks or months among a community in an effort to gain a deep understanding of local problems. In the next chapter, we will investigate how this kind of immersion can open up the imagination.

3

SOMEONE ELSE'S SHOES

I first heard about Amy Smith in 2003, when several friends told tales of an unorthodox MIT instructor who had moved her entire class to Haiti. In a reversal of the usual roles, the local farmers were educating the MIT kids about technology. This so intrigued me that I tracked down Smith—no easy matter, since she was often rushing to catch a flight to Ghana or Zambia. That's how I came to be jogging behind her one day, as she strode down an MIT hallway, carrying a bucket. She was on her way to the Charles River to retrieve some dirty water to use in a class demonstration. As she loped along, Smith described the ordeals of testing water in remote villages.

When we reached the massive front doors of MIT's main building, she pushed out into the crisp air and roar of traffic. Without interrupting her disquisition on water testing, she perched sidesaddle on a banister. Elegantly poised, she slid down the handrail, still talking. She landed on the sidewalk with a practiced leap. "You slide faster in the winter," she said, "when you've got a wool coat on." The banister was a perfect example of her design ethos: it required less energy than the stairs, and it was free.

• • •

In previous chapters, we've looked at how frustration, pain, and drudgery can inspire a technical insight. But all too often, engineers and designers are isolated from the conditions that would help them to understand other people's problems.

In fact, in many design schools, students are taught only to understand *their own* needs and think of *themselves* as the consumer. Left to their own devices, most college students will churn out the same "inventions" over and over again: devices that enhance the consumption of beer or phone apps that alert you to the presence of hotties. Inventor Mark Belinsky told me that "college students tend to target other college students. They're interested in dating and socializing. They're not trying to solve hunger, air quality, or health." In other words, many students never learn how to detect problems that affect people in other countries or who live in poverty.

Smith has devised her own system for teaching engineering as a form of empathy. This involves taking her entire class to live in an off-the-grid village and learn to appreciate the unique problems of that area. Together with local people, the students design devices that can help make life a bit easier.

When I sat in on Amy Smith's D-Lab class, the students gathered around a huge, black-topped slab of a table, chatting about their upcoming trip to Haiti—they would leave in a few weeks. In addition to helping local people solve technical problems, the students would also test village drinking water for dangerous bacteria. Smith was showing them how to do that—while putting her own ethical spin on the task.

She pointed to a piece of the water-testing rig—what looked like a silver barbell. "This test stand costs $600," she said. "Personally, I find that offensive." When the students worked in the field, she said, they'd be using a far cheaper setup, one that she patched together herself for about $20, using a Playtex baby bottle. "You can do a lot more testing for the same amount of money."

In the field, the students would have to place samples of the water in Petri dishes, and then would incubate the dishes for an entire day at a steady temperature. But how do you do that in a lean-to, with no electricity? Again, Smith had a solution. She passed around a mesh bag of what appeared to be white marbles. The "marbles" contained a chemical often used in polymer processing. When heated and kept in an insulated environment, these balls would hold at a steady thirty-seven degrees Celsius (human body temperature) for twenty-four hours. The balls were the crucial ingredient in one of Smith's inventions—a phase-change incubator that requires no electricity. (Called the PortaTherm, her off-the-grid incubator is currently being used to help diagnose typhoid and para-typhoid diseases in developing nations.)

Now, she wanted to demo a bacteria test in the dark, so she asked a student to cut the lights. He flipped a switch. For a moment, nothing happened, and then the room vibrated with a mechanical hum and panels closed over a bank of skylights in the ceiling. All craned their necks to watch. The panels moved in menacing slo-mo, like something out of a James Bond movie. A few people giggled, as if they had suddenly become aware of the ironies thrumming in this room—they'd come to one of the best-financed technical institutes in the world to learn how to work with baby bottles.

$2 a Day

Every year in her design lab, Smith gave her students a tough lesson in the gaps between first world and third. The students would spend a week surviving on $2 a day in Cambridge—the equivalent of what the average Haitian earned. One year, Jamy Drouillard, who was a teacher's assistant for Smith's class, performed the assignment along with his students. Drouillard had grown up in Haiti, but that didn't give him any advantage. He laughed, remembering his chief mistake.

"I bought a bunch of ramen noodles, a packet of hot dogs, a bunch of spaghetti, and some ketchup," he said. "It got sickening after Day 3. Actually, before Day 3. I should have mixed and matched instead of buying five boxes of spaghetti. In Haiti, people come up with creative ways of varying their food intake." He said the assignment drove home Smith's point quickly: living at subsistence level requires enormous creativity. The African farm woman who finds a way to make a scrap of land yield enough cassava root for her family is as much an innovator as any MIT-trained engineer.

Once, at an academic dinner with a plentiful buffet, Smith pulled out crackers from her pocket and nibbled while colleagues feasted around her; she was sticking to the $2-a-day assignment in fellowship with her students. It was relatively easy for her. She was living happily without kids, car, or retirement plan. Because she didn't see the point of finishing her PhD, she taught classes as an instructor at MIT—with the low pay to match. Her life was much like one of her own inventions, elegantly simple and off the grid. She'd first come up with her plan for dedicating herself to radical problem solving back in the 1980s as a young Peace Corps volunteer.

Back then, she spent hours pounding raw sorghum into flour in a remote village in Botswana; she lived among women who scrubbed, toted, lifted, and churned themselves into exhaustion. The drudgery of it became an education for her; she was outraged that these women had to grind their lives away.

Smith liked to fix machines and rewire radios. Her father taught semiconductor physics at MIT, and her mother was a math teacher; when she was a girl, her parents discussed the Pythagorean theorem at the dinner table. She had been raised to believe in the power of ingenious solutions.

One day in the 1980s, as she gazed out the window of her room at the expansive Kalahari pocked by thorn bushes, the arc of her life suddenly seemed obvious: she would earn a degree in engineering and then devote herself to improving conditions in this hardscrabble land. She intended to apply to both MIT and the University of Pennsylvania. But then fate intervened. "A cat had kittens on my grad applications for U. Penn. It was just covered in placenta stains, and I didn't feel I could send it in. So that's how I ended up returning to MIT," she told me. "That's how life works for me."

Sometime after she returned to Cambridge, Smith began trying to solve the problem that she had experienced in Botswana; she hoped to find a better way for people to process grain. At that time, some African villagers had access to a no-frills motorized grain mill, and when the machine worked, it could transform the lives of women. But there was a problem: the machine used a wire screen to sort the flour from the grit, and that screen tended to break, and it was nearly impossible to replace the part in a remote village in Ghana or India or Haiti. So as a result, the prized machine often ended up in a corner, gathering dust.

Smith had spent hours crouched in the dirt pounding at sorghum, so she knew how heartbreaking it would be to villagers when the machine failed. That inspired her to devise a mill that was much speedier and cheaper than its predecessor—and made entirely of parts that could be built by a local blacksmith. She used the airflow through the mill, rather than a metal wire, to separate the flour from the husk. She then worked her connections in the nonprofit world to manufacture and test the machine in Senegal, Haiti, and Ghana. Smith received acclaim for her grain mill, and she became the first woman ever to win the prestigious MIT Lemelson Student Prize for invention. She gives credit to her mentors in Africa—particularly the female farmers she met in Botswana—who educated her about ingenuity. "In Africa, the women are the farmers. Women invented domesticated crops. If you're talking to the right people, they should be a group of elderly women with their hair up in bandannas."

Smith, in turn, teaches her students to think about solutions that cost pennies and benefit thousands or millions of poor people. If you spend time around her, your ideas about invention begin to change.

"There are geniuses in Africa, but they're not getting the press," she told me, and gushed about Mohammed Bah Abba, a Nigerian teacher who came up with the pot-within-a-pot system. With nothing more than a big terra-cotta bowl, a little pot, some sand, and water, Abba created a refrigerator—the rig uses evaporation rather than electricity to keep vegetables cool. Her brand of invention requires humility as well as ingenuity, because your masterpiece may end up looking like a bunch of rocks or a pile of sand.

The Next Generation

Smith's students, most of them future engineers, seemed to absorb her lessons quickly; they think a lot about how technology can be applied to social problems. Instead of just asking whether the machine works, they wonder how it affects the economy of a village, or the environment. In 2003, I watched one of these students—Shawn Frayne, a gangly guy with a shock of black hair—work in a barbecue pit near the MIT student center. Pale blue smoke streamed out of a trash can and twisted in the direction of the tennis courts as Frayne stuck a lighter down into a can and tried to stoke the fire.

He held up one of his finished products—a piece of charcoal that looked like a jet-black hamburger patty. It was made out of the inedible parts of the sugar cane—that is, trash. These humble wads could help to solve a number of problems in Haiti: people could make their own charcoal rather than having to pay for the prefab variety, and that could in turn enrich local entrepreneurs; plus, the trash-charcoal would help to protect Haiti's endangered forests. Frayne had already graduated from MIT, but he was so devoted to Smith's design class that he'd stuck around to put finishing touches on several inventions. "I learned in an economics class that if someone has a good idea and they can implement it in a third-world country, they can dramatically change the economy of the country," Frayne told me. "I was surprised by how much technology can affect the well-being of a people."

A year after that, in 2004, he traveled to Haiti once again as a volunteer in the charcoal project, landing in a village with no electricity where local people struggled to pay for kerosene to light their houses. How, he wondered, would it be possible to create a power grid for the people here? And then one day, "I looked up and saw a flag whipping in the wind, and that gave me the idea," he later told me. He rigged up a tight belt of fabric and watched as it shivered with resonant energy in the wind, and this led him to dream up a new, low-cost method for producing power that could be fabricated in developing nations. His Windbelt generator—a pocket-size device that can be built for under $50—can power LED lights and radios. The technology won a *Popular Mechanics* Breakthrough Award in 2007, and a year later he was named one of the "Best Brains Under 40" by *Discover* magazine. After spending time in an off-the-grid village, Frayne had approached the energy problem from a new angle—one that he probably never would have grasped in the comfort of a first-world lab with plentiful electricity.

Inventor as Ethnographer

This kind of ethnographic fieldwork, in which the inventor spends time immersed in other people's problems, can be key to corporate breakthroughs too. During the 2000s, a team of academics (Abbie Griffin, Raymond Price, and Bruce Vojak) sought out star inventors at corporations like HP and Procter & Gamble. The researchers interviewed dozens of these high performers—whom they called Serial Innovators—and asked them how they had hatched ideas that led to million-dollar products. Many of these inventors reported that they spent as much time as possible among the customers they served. In some cases, the inventors spent days mingling with users at sites like hospitals or farms so that they could observe people and ask them questions like "Why are you doing that?"

"Serial Innovators have to live [inside the problem] and develop their own sense about what is important through firsthand experiences," according to the researchers. The Serial Innovators go to great lengths to learn about other people's pain and frustration—and their deep commitment to the customers pushes them to make sure that they are satisfying a true need.

"A gizmo might be clever or cool, but if no one wants it, then it's not an invention," Frayne said. "It only becomes an invention when someone cares."

Martin Cooper, developer of the first hand-held cell phone, echoed that thought. "That term, *useful*, is really . . . powerful . . . because what an inventor has to do is put his mind inside the mind of the user, not his own mind. [A device] may or may not be useful to the inventor, but it's got to be useful to some segment of society," he told me.

So inventors must be able to do more than just collect feedback; they also have to be able to listen to criticism and change course when they discover that they're solving the wrong problem. It takes enormous self-control to do that. Rather than falling in love with your ideas, you must seek out criticism and cultivate people who will squash your enthusiasm. When someone tells you, "I don't need your idea," you must not fling your drink in his or her face; instead, you must ask, "Why?" It's the most difficult—and ego-bruising—part of the creative process.

In the next chapter, we will look at how inventors gather and learn from feedback.

4

THE FUTURE OF FEEDBACK

In the early 1980s, Dick Belanger cofounded a successful company called Adhesive Technologies that made hot-glue guns—but soon he wanted out. The thought of spending the rest of his creative life in glue was soul crushing. Instead, he dreamed about whimsical inventions and even built some of them on weekends. For instance, he combined parts from a vacuum cleaner with grooming clippers to create a "hair-cutting machine"; as the clippers snipped, the hair flew into a box. And when his bathroom mirror fogged up, Belanger was inspired to invent a fogless mirror. Then there was the tennis ball reinflater. And the aluminum hockey stick. And the car engine pre-oiler. In the 1980s, he was generating so many concepts that he decided to begin collecting them in a notebook he titled "Dick's Book of Dumb Ideas"; within a few years, he had filled it with hundreds of entries.

He hoped that his ticket out of glue guns would come from one of the concepts in his notebook. "I actively started looking for the right idea to take a leap with," he told me. "I wanted to make one big deal with one company that would license my patent and then sell the heck out of it."

But as he considered all of the products that he'd sketched and prototyped over the years, he recognized a hard truth. He wasn't willing to stake his career on any of the products he had invented so far. For instance, he'd built a prototype of a heated mirror, hung it in his bathroom, and showed off its defogging action to friends and family. But how many consumers would go to the trouble of installing a brand-new bathroom mirror just to solve a minor annoyance? Much as he loved the gadget, he knew he couldn't afford to bet on it.

Nowadays, inventors can raise money on a crowdfunding site and then sell the product directly to customers; but back then, if independent inventors wanted to pursue an idea, they usually had to stake their life savings. In the 1980s, Belanger felt he had no choice but to try to license his concept to a large company. And preparing to negotiate with a company was expensive: he would first have to hire a lawyer, craft a professional tool-and-die model for factory production, and then pay for marketing materials—an investment of about $50,000. Of course, there was no guarantee that any company would want to buy his invention, so he would have to gamble with his family's savings.

That's why Belanger bided his time and kept adding entries to "Dick's Book of Dumb Ideas."

He could bet on only one idea, and it had to be a product that tapped into the needs of a large audience.

By the late 1980s, he had become a father. "I was right in the middle of it," he said of the child-care duties he shared with his wife. "I had no problem with diapers, but the spills drove me crazy." Like most parents at that time, he bought a cup with a snap-on lid that was supposed to thwart toddlers. His son Bryan delighted in foiling the contraption, shaking it upside down. One day in 1988, as Belanger cleaned up another spill, he thought, "I'm going to see if I can make this thing so it doesn't leak at all and outsmart the child."

Since he had lots of experience with glue guns, he knew what makes a nozzle work. He put together a prototype cup made out of Tupperware parts and added a built-in mouthpiece, experimenting with different kinds of valves until he found one combination that would let in some air as the child sipped. The vacuum trapped the liquid inside the cup—even when it was upside down.

Now Belanger decided to go forward with the idea in a modest way; he invested in the patent and the manufacture of several thousand sippy cups. For a couple of years, he and his family ran a cottage industry, selling the cups to friends and acquaintances. They even created just-for-fun "commercials" in which Bryan (and sometimes his younger brother, Steve) hammed it up as they demonstrated the product. The sippy cups sold like crazy. Before the advent of Internet crowdfunding, Belanger had found his own crowd. He'd connected with a community of parents and discovered that his customers were wildly enthusiastic about the sippy cup.

Belanger finally leaped. In the early nineties, he paid thousands of dollars to a patent lawyer who would help him draw up documents and enter negotiations. After a few tense weeks, Belanger signed a licensing deal with Playtex. His gamble had paid off; the deal was lucrative enough to support his family for years to come.

Playtex issued the sippy cups in pink and blue, and soon they were everywhere. On vacation at Disney World, the Belanger kids ran around, peering into strollers, counting up sippies, and calculating how much money "they" had made.

Belanger still has his notebook from the 1980s and sometimes he flips through it, thinking about all the products that he could have developed. Several of his concepts have since become commercial products—marketed by someone else. "Some entries that I had forgotten about showed up thirty years later," he told me. He feels a twinge of regret, of course, but he also knows that if he'd followed up, say, the tennis ball reinflator, he might have ended up regretting it. That word *dumb* on the cover of his notebook was smart. It helped him distance himself from his notions before he could grow too attached to them. It had been painful to let go of many of those projects, but he had weighed his regret against his fears about the future. What if he'd gone with the wrong idea? What if he'd lost his investment? The secret of Belanger's success had been his exquisite sensitivity to signals, both from his own gut and from his community.

The Pre-Mortem

The psychologist Gary Klein has spent decades studying people who perform at exceptional levels—from firefighters who escape from burning houses to scientists who crack open mysteries. He has found that the most skillful people continually replay their own mistakes in their minds. Even the tiniest error "still bothers them; they're still frustrated by it and wondering what they did wrong or how they can improve the process," he told me. The "experts have that mindset of continuous improvement."

According to Klein, if you ask beginners to describe their mistakes, they say, "I can't think of any." Some people are oblivious to the problems around them. And other people lie awake at three in the morning, obsessing about that one bolt that might work itself loose from the machine.

That's why the most successful people learn from incidents that the rest of us wouldn't even call mistakes. The hairs stand up on the back of their necks when something odd happens—they anticipate trouble in the pattern of clouds, a puff of smoke, a chance remark, a leak, a dimple in the road.

In other words, learning from feedback *requires the ability to feel imaginary pain.* You must be able to play out different scenarios in your mind and become terrified about all the things that could go wrong. And it also requires a unique kind of sensitivity, the ability to take instruction from failure and from people who tell us that we're wrong.

Klein has devised a method—he calls it the "pre-mortem"—to sharpen the imagination, using it as a tool to predict disaster. He asks executives to use their minds to time-travel into the future and then look *back* at the plan that they're about to put into action.

As Klein explains, "A typical pre-mortem begins after the team has been briefed on the plan. The leader starts the exercise by informing everyone that the project has failed spectacularly. Over the next few minutes those in the room independently write down every reason they can think of for the failure—especially the kinds of things they ordinarily wouldn't mention as potential problems."

One by one, people read off the reasons why they imagine that the project could spin out into a massive failure. The pre-mortem "reduces the kind of damn-the-torpedoes attitude often assumed by people who are over-invested in a project," wrote Klein. "The exercise also sensitizes the team to pick up early signs of trouble once the project gets under way. In the end, a pre-mortem may be the best way to circumvent any need for a painful postmortem."

If you're excited about a project, it can be hard to listen to your own doubts and worries—or even formulate them. There's an I'll-just-close-my-eyes-and-plunge exuberance that comes with a creative breakthrough. The last thing you want to do is honestly assess the outcome—which is why the pre-mortem can be such an effective blast of cold water. Belanger had intuitively used this method as he resisted the urge to go all-in on his fogless mirror or hair-cutting machine. He had "product-tested" his ideas inside his mind and *felt the pain* of the imaginary failure.

Many stories of invention involve an outlier who clung to his dream, ignored the experts, and prevailed against those who called him a crackpot. It may appear that the inventor succeeded by shutting out feedback. But when you investigate these cases closely, you almost always find out that the inventor was deeply interested in other people and their desires. Instead of paying attention to the conventional wisdom, the inventor was observing *different and better* feedback than everyone else.

Crowdfunding as Pre-Mortem

Back in the "old days" before crowdfunding, it was difficult to collect that feedback early in the design process. Inventors tested out their prototypes on friends and relatives, or they relied on their own gut instincts, or they performed pre-mortems in their imagination. Corporations often sought out feedback only late in the development cycle—for instance, by test-marketing a product that had already been manufactured.

But on a crowdfunding site like Kickstarter or Indiegogo, everything about market research has been reversed. Instead of *paying* people to give you feedback on your product, your focus group *pays you.* In a crowdfunding world, the audience can green-light a project or kill it. If you've asked for $50,000 on Kickstarter and your backers cough up only $45,000, the project is dissolved and you receive no money at all. Other sites abide by different rules, of course. But every flavor of crowdfunding comes with some kind of built-in pre-mortem. You're gathering customers *before* you go forward with your project, and you have no choice but to listen to them. The exercise that Gary Klein suggests—imagining failure before it happens—is no longer as crucial, because you have a chance to fail so early. And instead of imagining how people will respond to an idea, you can find out right away.

Of course, the Internet offers many strategies for close collaboration between inventors and users. As we saw in Chapter 1, even though Twitter was *not* launched on a crowdfunding site, it benefited from a community of obstreperous users who made their needs known. This "bucket brigade" strategy—in which users nurture a company that is creating a tool they want—can evolve in a number of ways. But crowdfunding seems to be particularly helpful for encouraging a pre-mortem mindset, because fans give feedback in the form of both suggestions *and* money.

Crowd-Whispering

Shawn Frayne, whom we met in Chapter 3, is now an independent inventor with a design shop in Hong Kong. When I asked him how he investigates the needs and desires of his audience, he answered with one word: "Kickstarter." Crowdfunding, he said, "is a hundred times more important than 3-D printing" or any of the other new tools that transform the way we make things. In recent years, he and his collaborators have made ample use of the Internet crowd to help them produce a string of new technologies, including a printer that can fabricate solar panels.

"Crowdfunding is an amazing way to get feedback from your potential users," Frayne said. "The first Kickstarter campaign that we ran, we asked for $25,000, and we had no idea whether anyone would want our product. We learned that people were so excited about the idea that we raised $150,000."

Though we tend to think of crowdfunding as a mechanism for raising money, Frayne told me that it's most valuable as a tool for reading the unexpressed desires of an audience. When you parade an early-stage project before the public, you have a chance to learn from users' suggestions and criticisms right from the beginning—and even to kill a doomed project before it sucks up years of your life. "It has always been insanely difficult to get immediate market feedback. But now you can get feedback as fast as you can create the prototype. In a way, Kickstarter is like rapid prototyping for market demand. You can throw it out to the crowd and immediately get an answer."

To give me an example of how he uses Kickstarter, Frayne told me about Looking Glass, a hologram technology that he recently developed with his business partner. A Looking Glass is a chunk of Lucite-like material that appears to contain a human heart, a pistol, a frog, a foot, or a tiny replica of your friend Bob. As you turn the Looking Glass around, you can view Bob (for example) from many angles; you see the rip in his T-shirt and the pompom on the top of his jaunty chapeau and the wrinkles around his eyes. The ten-inch-high Bob appears to be very real. But in fact, this is just a vivid illusion made from ink embedded in many layers of plastic. The technology was remarkable, and Frayne suspected that lots of people would want to use it for . . . something. But who would crave a mirage inside a plastic cube? What was the Looking Glass meant to *do*? How would it live in the world?

Frayne and his collaborators decided to run three or four campaigns on Kickstarter in which they would explain the Looking Glass technology, raise money, and—most important—learn from other people how to develop their product. "We call it 'micro-Kickstarting,'" he told me. With the first micro-Kickstart, they began conversations with several funders. "A biologist wants to make huge Looking Glass hologram crabs, and someone from the next World's Fair in Italy wants to use them in displays. These are people I never knew existed two weeks ago. So now, because of the micro-Kickstarting, we're able to match what we're making to people's desires," Frayne said. These strangers alerted him to their own needs.

In 2013, I interviewed another young inventor named Mark Belinsky, who was sipping coffee in a hotel room in Shenzhen, China, preparing for a day of touring factories. He and his partner had developed a product called Birdi that can sniff invisible dangers—from toxic chemicals to smog to smoke—and then send a report to your phone. Belinsky had dreamed up the idea only about nine months before I talked to him—and now he was in negotiations to produce it. This is an eye blink when you compare it to the ordeal that independent inventors faced in the 1980s.

When Belinsky and his friends needed seed money, they made their appeal on a crowdfunding site (Indiegogo), and in about two months they had raised more than $70,000.

Like Frayne, Belinsky discovered that the feedback could be as valuable as the funding. "It was really exciting to see a community spring up around the idea and push it in new directions," he said. For instance, one guy from Oklahoma wanted the Birdi to warn him about incoming tornadoes. "I'm from New York City, so I don't really think about tornadoes ever. But someone in Oklahoma thinks about this all the time, so of course that makes a lot of sense. That helped us to think about alerting people to all sorts of emergencies. A lot of our key insights came from crowdfunding."

Belinsky's method is a little different from Frayne's—he scours the messages not just to understand who wants the product, but also for the ideas (even inventive tweaks) that he can harvest from his crowd. This suggests that an entirely new kind of collaboration is afoot, one in which the inventor acts as a sort of ombudsman or curator.

In 2014, researchers at Northwestern University published a study that examined what happened when people tried to launch projects on Kickstarter and then failed to meet their funding goals. After all, there are lots of failures on crowdfunding sites—58 percent of creators are told no by the crowd on Kickstarter. The researchers, curious to know whether people learn from crowdfunding feedback, culled data from sixteen thousand Kickstarter campaigns and interviewed eleven project creators. "It appears that the creators of unsuccessful projects that do end up relaunching [the campaign] learn from the mistakes of their original projects and revise appropriately to become successful," the researchers wrote. The relaunched projects netted nearly $1,400 more (on average) than the first-time efforts. "I don't want to be cheesy but [failing on Kickstarter] made me stronger," according to one of the interviewees in the study.

Low-Cost Failure, Lots of Feedback

Chris Hawker believes that crowdfunding is especially useful because it lets you fail rapidly and cheaply. Hawker is an award-winning designer who has launched several of his own products on and off on crowdfunding sites. He also runs a firm called Trident Design, offering expertise to inventors who want to go forward with their own product ideas on Kickstarter and Indiegogo. When I talked to Hawker in 2014, he and his team had just helped a client to redesign what was then the most successful campaign in the history of Kickstarter. The client's product, a high-tech drinks cooler called the Coolest Cooler, drew 62,642 backers and more than $13 million.

Even coming off of this success, Hawker points out that most crowdfunding campaigns are destined to fail. "It's so hard to predict what's going to ring the bell. Only about one in eight of our products makes a lot of money," he said. And it's the failures that fascinate him. Whenever the crowd rejects a product, Hawker performs a forensic analysis to find out why customers said no. He often begins with a soul-searching conversation inside his own mind, and then with his clients and his employees. He reeled off the questions he poses to himself and his clients: "What's in the gap between the desired outcome and the outcome we got? Was it in the product design? Was it in the execution? Was it in the marketing? Was it in the cost? Was it in the presentation? What didn't work, and how can we learn from that for next time?"

As Hawker points out, the reasons why an audience connects with a new product are often intangible and mysterious. You can't reduce it to a formula or a method. Instead, you have to know the problem inside and out; you have to be able to think like the customers, to vibrate with their desires. "You've got to discern which is the right feedback and which is the wrong feedback. That's very challenging," Hawker said.

He told me that customers respond to some intangible quality in the design. The product must be more than functional; it must have a visual appeal that makes us crave it. That's why he takes the product through dozens of drafts, pushing for gradual improvement. When the design is right, he can feel it in his gut—and his eyes. He calls it an "eye-gasm." "People have to look at the product and think, 'Whoa, I've got to have that.'"

Hawker points out that if you try to pander to the crowd or win a popularity contest, you'll produce a cliché—a boring product that no one wants. Indeed, many companies have found that when they hand out surveys to customers asking them to express their needs, the results can be underwhelming. "If I had asked my customers what they wanted, they would have said a faster horse"—so goes a quote attributed to Henry Ford. Still, even if most people can only imagine faster horses, a few members of the crowd are the bellwethers and soothsayers. They're the Lead Users who have discovered a hidden problem that could inspire a hit product. The trouble is, it can be hard to know *which* users are the ones with the valuable ideas. Inventors like Shawn Frayne seek out the Lead Users and cultivate them to discover hidden markets.

But Hawker prefers to use the crowdfunding campaign as a way to observe how consumers will behave, rather than as a method to search for Lead Users. After all, what happens if people tell you they love an idea but refuse to pony up $20 for it? The best way to find out what customers will buy is to test the market early on in the design process.

Fifteen years ago, that was nearly impossible to do. In the late 1990s and early 2000s, when he began his career as an independent inventor, Hawker had to gamble his own savings and years of effort on each idea. He would develop the product to the point where it was ready to send to a factory for mass manufacturing and distribution—which could cost hundreds of thousands of dollars. It also took a long time to do market testing—at least "five years of waiting and investing and hoping to find out" whether the project would succeed, Hawker said.

Now, instead of five years it might take five months to find out how an audience will react. "At the end of the day, if you're an inventor, you're going into the unknown," Hawker pointed out, and risk is your business. That's why he believes that crowdfunding is so significant. The funders of an idea—rather than the inventor—take on much of the risk. At the same time, it radically speeds up the rate at which you can perform experiments that reveal consumers' cravings and desires. Crowdfunding, he said, means "we don't have to give up equity, and we don't have to go out and hunt down investors, and we don't have to take on any debt. It puts the power in the hands of the inventor and the consumers."

At the same time, he acknowledged, the crowdfunding world is still a Wild West teeming with scammers, incompetents, and pretend inventors who abscond with the money. The too-good-to-be-true technologies that set off a frenzy on crowdfunding sites often turn out to be just that. The would-be inventor is not skilled enough to manufacture the device and ends up frittering away the funds.

When you open up the R&D system to the crowd, "it's messy," Hawker said. "But that's the price we have to pay for shaking up industries and pushing them to reinvent themselves."

In this first part of the book, we have zeroed in on the kinds of inventions that emerge from a new need. But many problems don't have to be found, because we're all too aware of them. You don't have to launch a market research campaign to figure out that a safe, effective weight-loss pill would become a smash hit. The same applies to a cure for lung cancer. Or a battery that's so small, cheap, and powerful that we could use it to kick off a solar-power revolution. These problems aren't hiding; they're out in the open, taunting us, bedeviling us, tormenting us. We can't solve them because we're stumped about how to wrangle *stuff*—lithium, glass, human cells—into the shape of our desires. And we long for discoveries that will crack the mysteries.

In the next part of the book, we will look at how inventors hit on the key discoveries that open up new possibilities.

PART II

DISCOVERY

5

SUPER-ENCOUNTERERS

In 1982, a NASA engineer named Lonnie Johnson was tinkering at home. While working on a heat pump that would use water instead of Freon, he sculpted a nozzle to use in the device. When he attached the nozzle to a faucet to test it out, water shot across the bathroom with such force that it generated a whoosh of wind and hit the bathtub with a marvelous *splat*. Johnson was thrilled by the effect: "I've used plenty of hoses in my life, and they never inspired me to think about a toy. But the nozzle—that was different," he told me. The cartoonish action of the water inspired him, and he couldn't stop thinking about how this discovery could transform the way kids play with squirt guns. He saw the makings of a hit product.

Johnson spent years trying to make other people see that nozzle's potential by creating prototypes of a water gun. Wandering around at toy fairs with a suitcase under his arm, he hoped to find someone who would buy his idea. Eventually, a company called Larami licensed his water gun and sent it out to stores. Johnson's invention, the Super Soaker, did indeed become one of the blockbuster hits of the 1990s.

His idea emerged from a serendipitous event. He didn't go looking for the nozzle. Instead, it was as if the nozzle found him. This process is so common that we have a name for it: accidental invention.

In Part I of this book, we saw that people with a deep understanding of a problem can often figure out the best solution for it. Now, in Part II, we will turn our attention to inventors who operate in exactly the opposite manner. They stumble across a "preexisting solution," and then they work backward to figure out how to match it to a need. The microwave oven, Teflon, Velcro, the pacemaker, safety glass, x-rays—these all began when a lab worker bumped into some unusual phenomenon, became fascinated with it, and then figured out how to put it to use. These inventors often begin with a hunch that they've hit on a major discovery, but they may not know *why*—until years later.

When I interviewed Johnson, I was struck by his passion for that nozzle. It reminded me of the way someone might describe falling in love at first sight. Other inventors I'd interviewed had also described the same kind of powerful moment of recognition: "I saw it and I *just knew* it had to be important," they might say. If you look through the annals of technology, you can find hundreds of meet-cute stories, where someone becomes smitten with a discovery.

Indeed, some inventions could only have started with a random encounter—no principle or reasoning could have foretold their existence. In 1965, chemists at Searle were searching for a drug to treat ulcers; during these experiments, one lab worker, James Schlatter, was heating up aspartame in a flask when he spilled some of the compound on his hand. Later, when he happened to lick his finger, a glorious sweetness bloomed on his tongue. "I thought that I must have still had some sugar on my hands from earlier in the day," Schlatter wrote. But he eventually traced the mysterious sweetness back to that flask in which he'd been trying to brew a medication. As Searle scientist Robert H. Mazur notes, "The sweetness of aspartame could not have been predicted" ahead of time, because it was made from chemicals that were tasteless, sour, and bitter. In other words, the magic happened only with just the right formula—one that you would not begin to guess at from the drab flavors of the component parts. The historian Walter Gratzer notes that modern chemical sweeteners emerged in a volley of spills and pratfalls. For instance, in 1976 a professor asked his student to test the compound trichlorosucrose, later known as sucralose; the student, hearing the word *taste* instead of *test*, put it on his tongue.

"Where the frontier of knowledge is wide open, where causes of phenomena mostly hide in the dark, where many scientists deem research most challenging and exhilarating, chance and serendipity play bigger roles," according to Sunny Auyang, a philosopher who studies breakthroughs in science. Of course, when this creative process seems like an accident, we tend to treat it as an unusual event, even though thousands of major inventions have involved some type of serendipity.

In a 2005 European study of thousands of inventors, about half of them reported that their breakthroughs began with a surprise or an unsought discovery. Thirty-four percent of the patent holders had been working at their day job (which was not inventing) when they noticed a phenomenon or had a thought that inspired their breakthrough. Another 12 percent reported that their invention was "an unexpected byproduct" of their research. These figures make it clear that open-ended research is essential to many creative leaps.

In 1963, Duane Pearsall was working on a machine he called a "static neutralizer"; his idea was to dampen the static electricity in factories and photography labs, which could be annoying and potentially dangerous. One day, someone lit a cigarette in Pearsall's workroom and the fumes drifted through the air; that's when the antistatic meter began to spin crazily. Pearsall realized that the device must be reacting to the particles released by the burning cigarette. For a few days he was entranced by the effect, but he didn't see how it could be useful.

Then Pearsall ran into a friend who worked at Honeywell and

showed him the machine that could sniff the air for cigarettes. The friend offered some valuable advice: "Cut the static crap and develop a smoke detector."

At that time, house fires killed thousands of Americans every year. Though rudimentary smoke detectors did exist, they were expensive and unreliable and were rarely used in homes. Pearsall now understood that he had found the answer to one of the most important problems of his time.

He had no background in fire protection. He knew little about smoke-alarm engineering. Nonetheless, he spent years working on prototypes until he had developed the first low-cost, battery-powered smoke detector for the American home, launching a billion-dollar industry that would save dozens of lives every year. "We embarked on this experimental procedure without any forethought that the outcome would dramatically change the world. But it did," he wrote later. When he started out, he had been working to solve a minor problem, but the cigarette-sniffing device had inspired him to recognize a much more urgent social need.

Pearsall's story points us to a question: In the early 1960s, dozens of engineers were trying to invent a better fire alarm, so why did the experts miss, while Pearsall succeeded? Many of those engineers were on the wrong track: they believed that the best way to detect fire was to sense *heat.* For instance, an inventor named John Lindberg patented several heat-sensing alarms during the 1960s, each one an attempt to overcome the many flaws of this technology. He was still trying to work out the kinks in heat detectors when Pearsall released his smoke detector in 1968. Lindberg seems to have been an ingenious mechanic who cared deeply about saving lives, but he had traveled down the wrong path—fascinated with heat, he ignored the smoke.

So it may have been that Pearsall's outsider status gave him an edge; in fact, he knew so little about the problem that he relied on his friend from Honeywell to suggest a practical use for the cigarette-sniffing meter. At the same time, Pearsall himself must have possessed some quality—an unusual kind of imagination—that allowed him to become so passionate about the meter that he committed his life to it. Like Lonnie Johnson, Pearsall fell in love with his discovery.

Is "Luck" a Form of Creativity?

In the 1920s, Graham Wallas, a professor at the London School of Economics, devised a theory about how people think their way through creative puzzles and achieve breakthroughs. The process starts, he wrote, with the "Preparation Stage," when you find an interesting problem and decide to tackle it, then focus your attention on the mystery. As Wallas describes it, this first phase often involves conscious and effortful thinking, and you may feel stumped or blocked as you turn the problem around in your head. Next, in the "Incubation Stage," you knock off for a cigar, a nap, or a soak in the tub. During this resting period, your mind is free to roam and make new connections, and then revelation pops into your thoughts, seemingly without effort.

But there's a problem with Wallas's story about the birth of ideas: his model leaves little room for the role of discovery and serendipity. While Wallas's theory may describe the way some people engage in creative work, it can't be applied universally. As we've seen in this chapter, many inventors don't experience Wallas's Preparation Stage; they may not even be aware of a problem when they stumble across its solution. Indeed, starting with a question can sometimes *impede* creativity.

Steve Hollinger, a serial inventor who has developed everything from a large-format printing system to a shower drain, spends a lot of time just playing. For instance, once he started tossing an old camera across the room, and that led him to develop a throwable video camera that could film as it flew through the air. "You're feeling your way in the dark, and suddenly you discover something you didn't expect, and you think, 'Oh, my God, this is amazing.' You're working all the time; it's not a matter of an 'aha' moment or light bulb going on, because you're always engaged in experimenting. You have to take the time to try things and to be open to what happens."

Steve is a longtime friend of mine. When I started working on this book, I asked him if I could observe his creative method over a few months, so I could get a sense of how he juggles his many projects.

"So here's a weird thing I did yesterday," he said one night, as he drove me home from a restaurant where I'd been grilling him about his techniques. He had just cleared the passenger seat of old pens so that I could sit down, and that reminded him of his Sharpie experiment.

"I had this dried-up felt pen," he told me. "It seemed like a waste to throw it out, because there was good ink trapped inside it. So I thought, 'What if I put the pen in the microwave?' If I heated it up, some of the liquid inside the Sharpie would turn to steam, and that steam would push its way through the tip. So I threw the pen in the microwave for three seconds, and damned if that didn't work." He had become animated as he talked, his hands flying off the steering wheel. His method for re-animating pens seemed to work much better than the old lick-the-tip trick; he'd actually found a way to make them as good as new.

"What are you going to do with that idea?" I asked.

"Nothing," he told me, as he negotiated the traffic. "But it just felt good to find a way to fix something that bugged me."

A couple of weeks later, I visited Steve's loft and asked him whether he was still zapping Sharpies in the microwave.

"Oh, that," he said, and turned away. "I don't want to talk about that." I was surprised by his dismay at the very mention of the project. He had his back to me as he rooted through a box.

"Please?" I persisted. "What happened with the Sharpies?"

After much coaxing, Steve told me that he'd realized that the microwave trick didn't work well enough. "The pens drew faint lines and then dried up again."

"You were so excited about it the last time I saw you," I said.

"I know," he said. "But now I hate talking about it, because it didn't work. Hey! Look at this." Now he was screwing together what turned out to be a kayak light made of molded plastic; he told me that when kayakers navigate in the dark, their lives depend on their lights, so he had decided to create the ultimate fail-safe version of the product. To do that, he had spent a year paddling around Boston Harbor, testing out various companies' lights and discovering their weaknesses.

Now he was affixing one of his pole-mounted lights on a metal eyelet on a wood block; he whacked it so that it bobbed and snapped back into position. He did this to show me how the light would stay affixed to the kayak even in white-water rapids. I realized that Steve was like a kayaker in the night, following the bright beam of his enthusiasm. He used emotions as much as intellect to guide his decisions.

When you discover something that surprises you, you're enchanted, bewitched, and tantalized. A new door swings open, revealing a velvety darkness that beckons you. It requires whimsy and passion to recognize a clue when it appears before you. The happy accident is housed in the branch office of the imagination, because you have to be able to see possibility hidden in a nozzle, a moldy dish, or the images produced by a camera flying through the air.

"People bump into puzzles all the time, but most simply ignore them, rushing on in their daily businesses," according to Auyang, the philosopher. She believes that being able to spot an important discovery is a skill unto itself. "Serendipity—the knack of finding things not sought for—has won many trophies," she wrote.

The Power of Encountering

In the 1990s, the British psychology professor Richard Wiseman began to suspect that people who feel "lucky" tend to be especially observant—and that their ability to scan their surroundings makes it easier for them to notice useful clues in their environment. To test out this theory, Wiseman came up with a clever experiment. He placed ads in newspapers, asking to hear from people who regarded themselves as either lucky or cursed.

He received hundreds of replies. At one end of the spectrum was a woman who had met her romantic partner "by chance" at a party—an event that she attributed to her great good fortune. At the other end, a stewardess had experienced a lot of bad luck on her airline flights—one plane was struck by lightning—and she believed she had somehow "caused" the disasters.

Wisemen asked all of these people, the blessed and the cursed, to come into his lab and perform a task. He handed them a newspaper and asked them to comb through its pages and count every photograph. Wiseman had designed the task so that highly observant people would discover a shortcut. On the second page of the newspaper, he had planted a message: "Stop counting—there are 43 photographs in this newspaper." On average, the lucky people were far more likely to notice this message, and so they could come up with the correct answer in just a few seconds. Meanwhile, the unlucky people missed the clue, and so they took about two minutes to count the photographs.

Wiseman's experiment revealed that our expectation about happy surprises may affect the way we perceive the world around us. His results correlate with the findings of another researcher in an entirely different field.

In the 1990s, Sanda Erdelez—a University of Missouri professor interested in library science—studied the happy accidents that people make when they're hunting for information. She interviewed more than a hundred people to determine how they gathered material as they worked on a project. Some people—Erdelez called them "Super-Encounterers"—reported that they bumped into lots of unexpected discoveries as they worked. The Super-Encounterers planned on being surprised and believed themselves to be endowed with a special kind of perception that helped them to stumble across clues. Meanwhile, the "Non-Encounterers" remained narrowly focused on their task, and they rarely veered off their path to investigate a mystery.

Erdelez observed that the Super-Encounterers *relished* the search; they so enjoyed pottering and investigating that they even took on other people's problems and would make discoveries on behalf of friends, relatives, and colleagues. It's as if these people "have channels for information perception that are more sensitive than the channels of other[s]," she wrote.

More recently, UK researchers have launched their own investigation into how people stumble across discoveries. Dr. Stephann Makri, a specialist in human-computer interaction at City University London, has interviewed creative professionals to learn whether they rely on serendipity and unsought opportunities. One of the interviewees, a photographer, called herself "the queen of serendipitous events" and reported that she spent a lot of time rambling through the city. "You have to leave your house. You have to walk down the street. You have to look up, and then it happens to you," she said. Meanwhile, a visual artist told Makri that he spends some time every day standing on the sidewalk with his eyes closed, challenging himself to listen and absorb the music of the street. In other words, Makri found that the Super-Encounterers are deeply interested in what they *don't know.*

Makri told me that as he learned from the creative people in his study, his own attitude toward serendipity changed profoundly. "I was a very different person before I started this work—more rigid with my boundaries and focused on my goals," he said. After interviewing the kings and queens of serendipity, he decided to embrace random walks and chance encounters. "These days, I'll have a chat with anyone about anything," he told me. "I'm much more open to taking risks and to jump on opportunities. I make time to explore opportunities, even when it seems they're unlikely to pay off. In short, I've been serendipitized!" Makri did seem to luxuriate in the moment, and he delighted in new ideas. He happily conversed with me for hours in an open-ended way, as if we were wandering around a city together.

Despite his serendipitization, Makri acknowledges that there are downsides to all of this whimsy. You could spend years exploring the unknown and have nothing to show for it. More to the point, it's very hard to convince someone to hire you to chat with strangers, gaze at the clouds, and wander unknown streets.

Of course, the Super-Encounterers love to forage and will spend hours doing it, even if no one's paying them, because the search is intrinsically rewarding to them. They're happy to make a discovery, but also content to meander. And, paradoxically, their very willingness to stop pursuing goals may lead them to success.

The Art of Luck

Mihaly Csikszentmihalyi is a towering figure in the world of positive psychology and one of the most famous psychologists to study human imagination. In the 1970s, he and his collaborator, Jacob Getzels, performed a groundbreaking study that illuminated the wheels and gears inside the creative mind. Csikszentmihalyi and Getzels followed a group of thirty-one students at the School of the Art Institute of Chicago in an attempt to unlock a mystery: Which of these students would succeed in the crushingly competitive art world? And who would fail? And why?

To find out, Csikszentmihalyi and Getzels devised an ingenious method that let them peer into the students' minds. The researchers outfitted a room with art supplies and then laid out a buffet of objects that the young artists could use as inspiration for a still-life drawing; the objects included a velvet hat, a bunch of grapes, a brass horn, an antique book, and a glass prism. Each student entered the room alone; there he was instructed to spend time examining and choosing objects before he drew them.

Csikszentmihalyi and Getzels timed each artist as he picked out objects, arranged them, and sat down to draw. This meant that the researchers could gather precise data about how long each student spent playing, fiddling, thinking, sketching, and carrying out a plan. When Csikszentmihalyi and Getzels examined the data, they realized that the students fell into two distinct groups. Some students grabbed a few objects—say, the hat and the grapes—and began drawing almost immediately. These students reported later that they had *only one concept* and stuck to it. "As soon as I saw the objects arranged I had the idea of the final design," one student reported. Instead of generating lots of ideas, he had spent most of the hour executing his initial plan.

The students in the other group pursued the opposite strategy. They spent much of their time handling many objects—peering through the glass prism, petting the velvet hat, and flipping through the pages of the book. When these students began sketching, they did so in an exploratory way, without a sense of a final outcome. Instead of coming up with one idea and then executing it, they spent the hour observing, discovering, and improvising.

"I let [the drawing] grow . . . I felt it alive until it was done," said one of the students.

"I just decided to stop at this point, although I could have gone on and covered the whole wall," said another.

Csikszentmihalyi and Getzels then asked several panels of judges to pick the best drawings. (The panels included professional artists, art instructors, expert critics, and lay people.) The judges tended to give low scores to the first group of students—the ones who had seized on one idea and then stuck with it. The judges lauded the designs produced by the students who had explored, touched, experimented, and scribbled.

Amazingly, this *one-hour* test seemed to predict what would happen to the students later in life. Seven years after the study, the students who spent the least time handling the objects had failed to woo admirers—some of them had sold so little of their work that they decided to give up painting entirely. Meanwhile, many of the serendipity-minded students had been able to support themselves as professional artists or art teachers.

Bottom line: The successful students spent far more time exposing themselves to new ideas. During the hour, they continually tested and explored, figuring out how the drawing would look this way and that way. The successful artists loved to tinker—and tinkering is an activity that we also associate with invention. Yet this highly productive pursuit doesn't get much respect.

Tinkering

In Europe during the Middle Ages, ragged fix-it men roamed from town to town, where they stopped to repair hoes, pitchforks, and knives with dollops of tin. The word *tinker* was hurled as a slur at these impecunious Irish wanderers, and even today it implies a shadowy sense of insult. The *Oxford Dictionaries* define tinkering as trying "to repair or improve something in a casual or desultory way, often to no useful effect."

Perhaps we fail to appreciate tinkering because it often involves dirty work with grease and scavenged parts. To do it properly, you must shove your hands deep into the guts of an engine; you need to touch, to listen, and even to taste. This is not white-collar work—if you're serious about it, you'll end up smudged and dusty. And that perhaps explains why tinkering is often regarded as a hobbyist or weekend activity, and not something that serious people undertake.

An inventor named Doc Edgerton did his best to change that, by blurring the line between applied science and tinkering. For decades, beginning in the 1930s, Edgerton presided over an MIT lab nicknamed Strobe Alley, where he dragged in boxes of wires, broken machines, empty bombshells, and buckets of seawater. He played with materials that other people considered garbage, and he *tried* to make mistakes. A student named Marty Klein wandered into Strobe Alley as an undergraduate in 1961 and fell in love with the smell of the place; it reminded him of the junk shops in lower Manhattan—the perfume of connectors, coils, and burned wires. Here, invention was not sterile; it was stinky.

Maybe that stink was the secret formula, because Edgerton was one of the most prodigious American inventors of the twentieth century. He transformed the strobe light from an obscure technology to a fixture of modern life. He made flashing light cheap and portable and found endless applications for it, from the airport runway to the office copy machine. Today, Edgerton is best known for the photographs he took. His images have become icons of the twentieth century: the drop of milk exploding into a crown, a bullet hovering beside an apple, an atomic blast caught the instant before it mushroomed, and hummingbirds frozen in mid-flight. His strobe photographs illustrated scientific phenomena in a way that was instantly understandable to millions of people. Later in his career he developed sonar tools that revolutionized marine archaeology, again using new ways of imaging to explore the unknown.

Edgerton became one of the great teachers of tinkering, which he believed to be a passion anyone could develop. He reconceived the technology lab as a cross between a playground and a trash heap, so you might call him the Father of the Hackerspace. His Strobe Alley was a forerunner of the public workshops and community makerspaces that are proliferating today. Doc kept a sheaf of postcards in his pocket, reproductions of his famous photographs with his phone number on the back. He handed these out to everyone he met, sometimes with an invitation to stop by the lab. The postcards were tickets to the world of Doc. All were welcome.

While working on my column for the *New York Times*, I interviewed another master tinkerer, Scott Burnham, who has an ear for wonderful mistakes. In the mid-1970s Burnham worked as Hippie in Charge of Technology—his official title—for a maker of electric guitar cables and accessories called Pro Co Sound. Back then, he spent a lot of time hunkered down in a basement workshop in Kalamazoo, Michigan. One day, as he was soldering parts, he picked up a resistor of the wrong size—this was his lucky mistake—and attached it to a circuit board. The machine began shrieking and moaning. And that's how he heard a new sound; it was beautiful, haunting, ugly, and full of soul. He immediately recognized this as a major discovery, and he built that sound into a distortion pedal he named the Rat. Attached to a guitar, the Rat turned every note into a blast of outrage. By the 1980s, Pro Co was shipping tens of thousands of Rats every year. Bands from Nirvana to Radiohead used the pedal; it adds a snarl to innumerable hit songs.

Burnham made his discovery by listening, touching, and fiddling. He used an artisanal process that involved the hands and the ears and the eyes. Tinkering is slow, in the same sense that "slow food" is slow—meaning that you have to spend a lot of time just taking in the smells and sounds, holding a conversation with the physical world, and waiting to see what happens.

A professor emerita of cognitive science at Georgia Institute of Technology, Nancy Nersessian studies invention in a deep and painstaking way, shadowing scientists for years at a time as they struggle to unlock mysteries; she is trying to witness problem solving "in the wild," as it actually happens in real life. For instance, Nersessian observed a lab where a team of scientists was struggling to build a "brain in a dish"—a collection of neurons that could control a robot arm. Nersessian watched the scientists as they played around with tissue, drew diagrams, programmed computer simulations, held meetings, and confronted failures.

After decades of conducting this kind of fieldwork, she has come to believe that one of the most important parts of the inventing process involves an unusual ability to visualize an imaginary world—to "prototype" an idea by constructing it in the mind's eye. After that mental feat, many inventors and scientists translate their idea into a physical model, like a sketch in a notebook or a computer simulation.

"The 'aha' moment is overrated," she told me. The real creativity and insight occur as people struggle with a problem in their minds, and then as they try to translate that design onto paper, a computer, PVC, Plexiglas, metal widgets, or living cell tissue. "The model often doesn't work, so they have to start the process all over again," she explained. "It's amazing how much failure people have to deal with on a daily basis. Things go wrong all the time." And when things do go wrong, you have to rethink your biases and assumptions. She observed that scientists engage in a dialogue with the physical world. They're continually surprised by results that they didn't anticipate, and sometimes they can transform an anomaly or an unexpected phenomenon into a brand-new solution. That's how they get lucky.

This luck, of course, can be enormously expensive—and it is why our government spends billions of dollars a year funding basic science, and why we should spend even more on it. Many inventive breakthroughs come from a *discovery of something truly new*—the kind of breakthrough that evolves out of painstaking experimentation or long years of fieldwork. Even then, the usefulness of such a breakthrough isn't always immediately clear.

In the early 1960s, "colleagues used to tease me about it," Charles H. Townes wrote of his pioneering work with lasers. At the time, the discovery struck many people as silly. Who needed a beam of light that burned a hole in the wall? His contemporaries mocked the laser as "a solution looking for a problem."

But decades later, the laser has revealed itself to be the solution to a plethora of problems—from eye surgery to computer-chip fabrication to fiber optics. "Many of today's practical technologies result from basic science done years to decades before," wrote Townes. "The people involved, motivated mainly by curiosity, often have little idea as to where their research will lead. . . . This springs from a simple truth: new ideas discovered in the process of research are really *new*." The very "uselessness" of a technology may act as a catalyst that mixes together diverse groups of people—it works like a spotlight, illuminating new problems and pushing us to imagine the impossible.

This is the reason we pay scientific researchers to undertake the examination of bird poop and dirt piles and polar ice; we do this because nothing beats random trials for turning up massive insights. At the same time, this kind of random investigation can be so expensive and risky that it discourages us from even attempting to find the big score out there in the darkness. For instance, it costs pharmaceutical companies more than $2 billion to develop a single new drug—and that cost scares away potential explorers.

But now, thanks to enormous storehouses of scientific data on the Internet, we may be able to radically reduce the cost of certain kinds of R&D. We're entering an era in which an ingenious grad student with a laptop might be able to use that data to make the next big discovery. This is why a new breed of inventor is tunneling through mountains of information to find happy accidents.

6

DATA GOGGLES

One of the most famous stories of medical serendipity involves a drug—originally called UK-92480—designed to open up blood vessels in patients with angina. While researchers were testing the compound on volunteers, they noticed something, er, exciting: men taking the drug reported that they'd had spectacular erections. "None of us at Pfizer thought much of this side effect at the time," according to inventor Ian Osterloh. But as the results came back, "we decided to follow up on these reports to see where it would take us." The discovery of the blockbuster drug that would become known as Viagra gives new meaning to the term *happy accident.*

Many of the great medicines of the twentieth century came about in just such a way—researchers were seeking *this* and they discovered *that.* Success was maddeningly elusive and terribly expensive; companies could aim hundreds of scientists at a goal and come up with zilch. And so several decades ago, pharmaceutical companies began to pursue what they called "targeted searches" or "rational design." Instead of waiting around for serendipity, you'd try to reason your way to a discovery—for instance, by identifying proteins that were thought to be involved in a disease, and then trying to engineer compounds that would affect those proteins.

In a 2013 paper published in *Clinical & Experimental Pharmacology*, a group of researchers pointed out that the rate of discovery of psychiatric drugs has been "anemic" in recent years. They raised a fascinating question: Why were the scientists of the 1950s and 1960s, using relatively primitive methods, able to unleash a dazzling array of new treatments for mental disorders? And why has the rate of discovery plummeted since then? "It is critically important to identify factors that may now stifle drug development," they wrote. "At least one prominent expert has attributed the apparent stagnation to a decline in serendipity."

The authors of this paper identify a whole range of what they call "anti-serendipity" factors that may be afflicting drug design. They point out that today many companies eschew completely random trials and experiments, embracing instead a more "rational" method. This involves "singling out for clinical trial only those compounds that basic research and theory suggest" will be valuable. In other words, rather than looking through the entire haystack, you search only the part where you believe the needle must be hidden. The authors argue that this targeted approach may reduce the chance of happy accidents. They also note that in today's hectic hospitals, clinicians spend less time observing and listening to patients, thus decreasing the chance that doctors will learn from afflicted people.

It may be that as drug companies spend billions of dollars trying to avoid fruitless investigation, they end up missing out on unexpected and surprising discoveries. It's like the joke about the guy on his hands and knees under the streetlight—he lost his keys in the dark, but he's decided to look where the light is better.

However, in just the past few years, a new breed of scientists—the bioinformatics people—have become excited about serendipity again. They hope to find a way to speed up and engineer their luck by using computers to scan the results of thousands of past experiments in order to detect unexpected connections. In theory, this tool would help them zero in on the kind of valuable insight that, for instance, transformed compound UK-92480 into Viagra—but they could do it much more quickly and cheaply. A 2013 McKinsey Global Institute report estimated that data-mining techniques could be worth $100 billion *every year* in the US health-care market—in part because the new tools might radically reduce the cost of pharmaceutical R&D.

"The time will soon come when big data is as much a part of both drug discovery and healthcare as it is of financial forecasting and choosing driving routes that minimize traffic," according to Steve Dickman, CEO of biotech consulting firm CBT Advisors.

But will this method really work? That question is bigger than big data: it forces us to grapple with the nature of human imagination and the ways that we explore the unknown.

As I was researching this question, I stumbled (serendipitously) across a story that reveals how the newfangled serendipity can be combined with old-fashioned detective work to make major breakthroughs. The story concerns a pharmaceutical scientist named Murray Robinson and a mystery that hid inside his own brother. The first clues to that mystery came from a clinician whose close observation of patients—and a hunch—led her to identify a problem. But the more complex nuances of the mystery will have to be cracked by tools like gene sequencers and data analytics, which allow researchers to explore with a speed and precision that would have been impossible even a decade ago.

In 1970, when Murray Robinson was eight years old, he shared a bedroom with his older brother, Kelly, and the two boys would whisper to each other after lights out, recapping their adventures from the day. Adults called Kelly "slow" because of his learning disabilities, but Murray thought of his brother as fast, loud, and funny. Kelly shoved his way to the head of the line at the ice cream store; he danced with his arms flailing and cracked fart jokes. In a hushed museum, Kelly pointed at a Greek statue and yelled, "Oh, my God, a naked lady!" In the dark movie theater, he called out, "Hey, everybody, let's play musical chairs!" And then there was his fascination with electronics: you had to hide flashlights and transistor radios; otherwise he would take the machines apart, strewing wires and batteries on the floor. Kelly was a florid eccentric, and Murray loved him for it. Years later, Murray would realize that all of these quirks were data points.

Fast-forward to the 1990s. At that time, Murray Robinson managed a research team for the biotech company Amgen, with a mission to develop new cancer drugs. "The genomics era had just begun," Robinson told me, and so "we jumped on it." Amgen bought state-of-the-art DNA sequencers, and Robinson and his team began to amass a library of genes involved in cancer. "It was the dawn of big data in biology, and we realized that we had to invent new ways of making sense of the data. I spent a lot of time with mathematicians trying to learn how to organize the genes into something that made sense."

Connections can remain entirely hidden for centuries before someone finds the right dots and links them. The clues are not "out there," inaccessible to us. They're all around us. And yet we don't pick them up. In 1998, as Murray Robinson was struggling to make sense of a blizzard of data, he got a call from his mother. She'd just had Kelly tested for a newly discovered genetic syndrome called Smith-Magenis. Though Robinson was an expert in sequencing genomes, it hadn't occurred to him until that moment that Kelly's behavior might have been shaped by a genetic mutation.

That day, Murray used Amgen's medical-search database to track down every study that then existed on Smith-Magenis. "I'm reading through one of the papers, and there's a list of all the traits associated with the syndrome—a gregarious personality, sense of humor, biting of their own hands, tantrums," remembers Robinson. "All the behaviors sounded just like Kelly. And then I got down to the bottom of the list, to the thing I found most shocking. It said that most of the people with this genetic mutation had a fascination with electronics. And my jaw dropped." Robinson remembered all those times when his brother had taken apart a light switch or pried apart a remote control to see what was inside of it. Now, Robinson struggled to reevaluate everything he knew about his brother's "personality." Kelly's passions, his obsessions, and even his sense of humor had all been shaped by a small error in his genetic material. (Kelly's appearance had also been affected by the gene; the condition had given him a heavy brow, cupid's-bow lips, and stubby fingers.)

Murray had always thought of his brother as unique, a real character, but now it turned out that Kelly belonged to a tribe scattered around the world, an invisible family, and they were all remarkably similar. It still surprises Murray that he—armed with the latest in genetic-sequencing machines—never wondered whether his brother's eccentricities could be explained by a mutation.

Smith-Magenis Syndrome was discovered the old-fashioned way, through careful observation and a good deal of luck. In 1981, a genetic counselor named Ann Smith examined an infant at a hospital in Denver, noting that the child had defects of the heart and palate. Months later, she treated another baby with almost exactly the same set of traits. Following a hunch, Smith suspected a connection and analyzed genetic material taken from the two children. She discovered that both infants were missing part of chromosome 17. Smith at first believed that she had simply found evidence of an *already-known problem.* So she hunted through the medical literature, searching for studies to confirm her observations. That's how she discovered that nothing had ever been published on this gene deletion.

Soon Smith began working with a genetic scientist—Ellen Magenis—to find and examine children born with the rare gene deletion on chromosome 17. By the end of the decade, the disorder had a name,

Smith-Magenis Syndrome, in honor of the two women who proved its existence. By then, researchers had discovered dozens of people with the gene deletion and had begun to catalog a long list of traits.

When Ann Smith first stumbled onto the mystery, she did not realize it was a mystery. She wasn't looking for the breakthrough. Instead, the discovery seemed to find her. Luck had a lot to do with it. Only about one in every twenty-five thousand people is born with the gene deletion, and Smith had the fortune to examine two rare cases within a few months of each other. If the children had been different ages, she might not have noticed the similarities between them, so that was another lucky break. And yet, Smith also seemed to possess some special talent that allowed her to see what others couldn't. She seized on just two cases, connecting just two dots to see a pattern. In Chapter 5, we learned that the term *Super-Encounterer* can be used to describe virtuosos who are prone to noticing patterns and making discoveries. Ann Smith certainly qualifies as a Super-Encounterer; she's an advertisement for the power of intuition, close observation, curiosity, and tireless detective work. Hers is the artisanal form of serendipity.

In 2009, Murray and Kelly Robinson drove to a Hyatt hotel in Reston, Virginia. Before stepping into the hotel that day, neither of them had ever met anyone else with Smith-Magenis Syndrome. But as soon as they passed through the doors of the conference room, they were surrounded by dozens of children—and a few adults—with the stubby fingers, heavy brows, and shambling gait caused by the gene deletion.

At that conference, Kelly didn't show much interest in the Smith-Magenis kids scooting around his legs. Instead, he stepped right up to a microphone in one of the conference rooms and began to examine how it connected to the cable—obsessed, as always, with a gadget. Meanwhile, Murray Robinson shook hands and studied name tags—he was reveling in the chance to talk to scientists about treatments that might help his brother. By then, researchers had established that the deletion of just one gene (RAI1) caused most of the problems associated with Smith-Magenis Syndrome. Still unanswered were the big questions. How did the gene work? What was it doing in the body? "I met clinicians and the physicians and the geneticists," Murray Robinson said. "One of the things that really struck me was how helpless they were at getting at the molecular root cause of this. And I thought, 'I can do something about that.' I realized that we must find a way to figure out what genes do," Robinson said. "That's the next big problem in my field. And so, because of my brother and because of what I'd been doing in cancer research, I decided to develop some tools to be able to interpret the genes." Though Smith-Magenis Syndrome begins with a single deletion, its effects touch many other genes, turning them on and off in ways that can be disastrous. Robinson wanted to see whether he could use data tools to find out more.

The technique is still so new that no one can yet agree on a name. Some call it "dry biology," while others prefer the term "bioinformatics," and still others call it "biology *in silico*." And it has deep implications for the way we will invent in the future. "An enormous amount of information—about genes, proteins, drugs, and diseases—is being dumped onto the Internet every day. It's a massive amount of new information, and most of it has never been analyzed," Robinson told me. "All this information is sitting there publicly available for anybody to analyze. Anyone can use it to make huge discoveries."

In order to investigate the RAI1 gene, and to understand the havoc that it caused in his brother's body, Robinson assembled a team: "I got together with some coders and they put a huge pile of data on a server in North Carolina, and we played around with different algorithms to try to link known genes together mathematically," he said. They found patterns in the data that suggest the Smith-Magenis deletion disrupts the way that DNA is packaged in the cell—which would explain why one broken gene can affect everything from sleep patterns to finger shape.

As he tinkered with the publicly available databases that contain information gleaned from thousands of trials and other studies, Robinson became excited by the possibilities. Like many other scientists, he believes that this kind of data mining may allow us to find new ways to deploy familiar drugs. The right algorithms could match the genetic problems caused by a disease with the actions of drugs and find a good fit. "Instead of waiting for the happy accidents in the lab, you might be able to find them in the data. You might be able to say, 'Hey, this drug that people tried for arthritis didn't work, but it actually might be a great treatment for hypertension,'" Robinson explained.

These days, with the explosion of the bioinformatics field, the hope is that we will find much cheaper ways to stumble across big discoveries. For instance, with mountains of data and just the right tools, you might be able to search thousands of random trials in an afternoon.

Robinson is betting his own future on it. In 2013, he founded a company called Molquant to find hidden insights in the data and translate them into drugs and diagnostic tools.

Engineered Serendipity?

Atul Butte, a biomedical researcher and entrepreneur who heads up the University of California San Francisco's Institute for Computational Health Sciences, is an enthusiastic booster of big-data techniques. At a 2014 conference, Butte pulled a gene chip (also known as a DNA microarray) from his pocket and brandished it in front of the audience. It didn't look like much—just a plastic square the size of a cracker. But, he explained, this device is transforming the way that we search the unknown.

When medical researchers use the chip to study how a certain drug affects blood pressure, for instance, they're also vacuuming up enormous amounts of data about how that drug interacts with all the other genes in the DNA sequence. In 2002, many medical journals began to require scientists to load all of this data from the gene-expression studies into public storehouses, even if only some of those results were described in the journal. That means the amount of surplus data—the information collected inadvertently during studies—is now so abundant that it's measured in petabytes. (Each petabyte is equivalent to 1,000,000,000,000,000 bytes, or about four times the amount of information stored in the Library of Congress.)

That's why some researchers, like Butte, are embracing a new strategy: instead of avoiding the unknown, they're finding ways to hunt through the unknown at top speed. They do this using algorithms to search through data from thousands of experiments that have already been performed in some of the top laboratories around the world—these results are pulled from data that researchers have collected but may never have used.

This data promises to transform our ability to discover new uses for familiar drugs. Earlier in this chapter, we saw that a medication developed for angina turned out to be surprisingly useful as a treatment for erectile dysfunction—but that discovery required a happy accident. "Historically, the discovery of new uses of old drugs is mostly through serendipity," wrote the authors of a 2015 paper. But nowadays, they point out, it's possible to speed up the happy accidents by finding patterns associated with the disease state (like gene expression) and matching these to the actions of drugs. Netflix analyzes its users' preferences to predict who will enjoy a new movie; Facebook mines networks of friends to figure out how to serve up ads; and now that kind of data analysis is moving into pharmacology.

Butte and his team have already begun mining public databases to hit on valuable new applications for old drugs. When they unleashed their algorithms, they discovered a pattern that suggested that imipramine (an antidepressant that has been around since the 1950s) could be effective against small-cell lung cancer. Butte and his colleagues then tested this theory in mouse studies. As he'd hoped, it turned out that the drug could be used to shrink cancer tumors in the lab animals. Butte and his colleagues published the results in the journal *Cancer Discovery;* "our work shows the power of bioinformatics-based drug approaches to rapidly repurpose FDA-approved drugs . . . to treat patients with [small-cell lung cancer], a cancer for which no effective novel systemic treatments have been identified in several decades," they wrote. The remarkable thing about this process is its speed: researchers moved from the clues in the data to clinical trials in about two years. If you had to wait around for serendipity, it might take decades to connect the dots.

Butte told me that his algorithms have alerted him to many other possibilities, including patterns that could lead to diagnostic tools for early detection of diseases: "We're looking at pancreatic cancer, skin disorders, and ulcerative colitis. And we have a new project to try to figure out preterm birth, or why do babies get born early, using public data that's funded by the March of Dimes." Through data mining, he and his colleagues found several proteins that give early warning of preeclampsia, a condition that threatens pregnant women and their unborn babies. The discovery of these biomarkers led to the development of an early-detection test that may be available by the time this book is published. The process happened at blistering speed—about two years from data mining to product development.

"Public data is like no other commodity in the world. It's not like oil, it's not like land, it's not like the air; it's certainly not like water. It's replenishable and usable by anyone. And if you kind of combine each data set with something different and ask the right questions, you get a new discovery," Butte told me. If you type the words *breast cancer* into one of the public data repositories, you will net tens of thousands of genetic sequences from breast cancer tumors—making this so accessible that even a high school kid can use it to engage in research science. Butte told me about a Duke undergraduate named Brittany Wenger who used public data to map breast cancer malignancies when she was just a teenager. In Chapter 16, we will meet another teenager who used online databases to devise an early-detection test for pancreatic cancer.

Butte, for his part, is discovering so much in the data that he sometimes feels overwhelmed. "We really have to get more people involved. There's too much for just me to do." And of course, when a lot more people are searching for drugs, there are a lot more chances for serendipitous discovery.

Big Pharma versus Little Pharma?

In previous chapters, we have seen that far more people are now participating in product design because they can get their hands on cheap, accessible R&D tools. Today, almost all of us have access to the kind of prototyping tools (like 3-D printers) that would have been available only to a handful of top engineers back in the 1980s. This trend is bringing a flood of new talent and new ideas into the realm of product design. Independent inventors like Shawn Frayne can now set up shop, control their own production, and rely on a crowd of backers.

For that same trend to sweep through the pharmaceutical industry, of course, the tools and the funding would have to be available to anyone with the talent to make use of them—and for the moment, we're very far from that situation. The pharmaceutical realm is still a gated community, limited to researchers who have access to labs, pathology samples, patients, and—most important—enormous amounts of capital. One way to speed up drug discovery would be to create affordable tools that would allow thousands of talented people to become, in effect, pharmaceutical companies.

And that's where we're headed.

It was Butte who first alerted me to a surprising new development: many of the facilities that you would need to undertake drug development are now available online. Just as it has become possible for garage inventors to collaborate with factories, it is also becoming possible for garage biotech to hire labs and other services crucial for proving the efficacy of a medical compound.

Under the conventional drug development system, researchers must budget millions of dollars to test compounds on animals in their affiliated labs. But there's now a work-around available on the Internet. An online site called Assay Depot lets you hire researchers to perform the studies of your own design on mice, rats, tissues, or cells. You can specify, for instance, exactly what kind of knockout mice you'd like to use to test your compound; these are mice that have been genetically engineered for obesity, lung cancer, leukemia, or other conditions. Even more intriguing, you can hire the labs affiliated with Assay Depot to create "designer mice" for your experiments. Anyone with a credit card can now buy many of the pieces needed to perform research, in the way that product designers can now order parts directly from factories in China.

Atul Butte has already begun using Assay Depot to test out the effectiveness of drugs. With a few clicks and a charge to his credit accounts, he has hired researchers at several well-regarded labs to perform the trials on drug candidates. Butte estimates that using this method costs his group about $75,000 (including his own labor) to perform a battery of lab experiments. And by taking advantage of the Internet, he estimates, he has been able to reduce the start-to-finish cost of researching a drug candidate to about $150,000. That price brings pharmaceutical R&D into the realm of possibility for even small operations, including Butte's own start-up company, NuMedii.

Butte told me that it's also possible to buy tissue samples online.

"The samples are stripped of patient identities, repackaged, and just waiting for researchers. So getting access to samples is not the hard part anymore, especially for commonly diagnosed conditions," he said. "You can actually get started right off the Internet."

Indeed, I found a company called Advanced Tissue Services that advertises itself as "your source of cells, human tissues and biospecimens." Though the website lacks a handy shopping-cart feature, it does promise that "fresh tissue may be obtained as per your specific requirements."

It's likely that some entrepreneurs will begin to figure out how to patch together these services into scrappy biotech companies. And then the Big Pharma companies may find themselves competing with a swarm of Little Pharmas.

"In Silicon Valley, we're so used to the kids today starting amazing, high-value companies in their garages and dorm rooms," Butte said. "Why couldn't you launch the next Genentech in someone's garage?"

I pointed out that because he's a university researcher with both a PhD and an MD, it's relatively easy for him to arrange clinical trials in humans. Kids in garages can't work those kinds of connections.

"You can't do it right now," he responded. "But why not in the future?"

In 2014, Y Combinator—the renowned start-up incubator that helped to create such hits as Airbnb—announced that it would begin investing in pharmaceutical companies.

One of the new Y Combinator ventures, Transcriptic, is building a service that will allow anyone to hire robots to perform automated experiments. "When we started Transcriptic, we set out with the goal of giving the life sciences the same structural advantages that [the] web has enjoyed, making it possible for two postdocs with a laptop in a coffee shop to run a drug company without the need for millions of dollars in capital equipment or lab space," founder Max Hodak wrote in a 2015 blog post. But, "to be clear, we are not there yet."

Right now, drug discovery is just in the beginning stages of opening up. Just as Uber has faced legal battles and pushback, so too will the for-hire labs. But it seems inevitable that some kinds of drug discovery will be performed on a laptop. And already bio-hackers are beginning to enter the field.

Still, Dr. Yogen Saunthararajah, a hematologist and oncologist at Cleveland Clinic hospital, has his doubts about engineered serendipity. For the past decade, he has been performing clinical trials to develop better medications for treating blood cancers, and in 2015 he and his colleagues published a breakthrough in *The Journal of Clinical Investigation.* Saunthararajah is a kindred spirit to Ann Smith, a Super-Encounterer who is deeply immersed in treating and listening to patients and dedicated to the time-honored craft of discovery. He told me he's skeptical about patterns plucked from a mountain of data and confirmed by experiments performed from afar.

Saunthararajah is deeply involved in the lives of patients with blood cancers, and he agonizes about the terrible side effects of traditional chemo treatments—exhaustion, hair loss, depression, vomiting, and even death. More than a decade ago, he began to wonder whether he could use low, nontoxic doses of an FDA-approved drug (decitabine) complemented by a non-FDA-approved drug (tetrahydrouridine) as an alternative to traditional chemo protocols. His hope was to find a more effective treatment that patients could undergo without sacrificing their quality of life. He theorized that these two drugs, taken together in just the right amounts, might activate a set of instructions inside the cancer cell that would make it "grow up" into a harmless blood cell and then eventually die. The medicine would have little effect on healthy cells, which means that it would cause few side effects.

In 2015, he and his colleagues reported that they had tested this theory on twenty-five patients with blood cancers. Nearly half of these patients saw improvements in their blood counts that signaled their bodies were fighting off the cancer. This was a particularly exciting result given that many of these patients were at the end stage of the disease and had not responded to other drugs. Just as important, the patients had experienced few of the terrible side effects usually associated with some kinds of chemo.

To achieve these results, Saunthararajah and his colleagues spent years in careful trial-and-error hypothesis-testing research. He emphasized that drug development requires deep involvement with both patients and science. Shortcuts rarely work. "To use a tired but valid cliché, correlation is not causation, and if we really want to make progress in fixing disease, we need to understand the chain of cause and effect in excruciating molecular detail," he said. "A robot is not going to reveal these mechanics, these truths. It takes artistry and love, and lots of funds, applied for decades. Hammering away at the kinks and details with human ingenuity and emotion does not make the news, but this is how the real job gets done."

But do we have to choose between the robots and the humans? The ideal solution would deploy both to discover new medical treatments. Many gifted researchers are finding ingenious ways to combine engineered serendipity and old-fashioned intuition to speed up breakthroughs.

After all, the $4 billion drug seems like a relic from another age, like the $1 million computer or the $2,000 calculator. The Internet has a way of removing barriers and giving outsiders the tools they need to challenge the incumbents. Already start-ups and independent inventors are beginning to scamper into drug R&D space and experiment with the new methods. And let's not forget those petabytes of data, like unexplored continents that hold the clues to human disease. "That data is frozen knowledge. Someone has to bring the heat to melt this knowledge, to melt this ice and let it out," Butte told me. Inside the frozen tundra of data may lie vast reserves of medical gold.

All the while, scientists are developing new technologies that let us search both the data and our environment for lifesaving drugs.

For instance, gene-sequencing tools have helped to reveal the hidden treasures in pond scum, soil, and the human gut; the microbes that would have been invisible to us a few decades ago turn out to be a promising new source of medicines. What once looked like nothing turns out to contain enormous value. In the next chapter, we will dive deeper into the glorious Dumpster of nothingness to find out why it is so inspiring to inventors.

7

BUILDING AN EMPIRE OUT OF NOTHING

In the 1980s, Jerry Seinfeld and his friend Larry David were wandering through a Korean deli, riffing about the snack foods on the shelves, when they realized that *this*—an ordinary moment in a comedian's day—could become the basis for a TV show. They wrote up a pitch and sent it to NBC. "That was the concept behind the show: no concept," Seinfeld remarked later. The NBC executives were dubious, so they stuck *The Seinfeld Chronicles* into a summer slot and waited for it to die.

But it didn't die. By 1992, *Seinfeld* was winning Emmys and on its way to becoming one of the top sitcoms of all time. By then (thanks to a line of dialogue uttered by the character George Costanza in one episode) *Seinfeld* had become known as the "show about nothing," because it dug deep into the nothingness of an Upper West Side comedian and his friends; once the characters spent an entire episode bickering while they waited for a table at a Chinese restaurant. The show zeroed in on life's in-between moments, finding an entirely new mine for jokes.

Some of the greatest discoveries involve a Seinfeld-ian paradox: enormous value hides in the realm that most people call "nothing."

Consider penicillin, which emerged from a dish of moldy goo that Alexander Fleming could have poured down the drain. The big ideas have a mischievous way of hiding in the trash. That's probably because when we label something as "waste" or "worthless" or "free," we stop paying attention to it. It takes a particular kind of imagination to see value in what others don't see at all.

A Fortune Built on One Piece of Paper

In the early 1960s, Lawrence Herbert glided to work in his navy-blue Cadillac. He loved his car, with its cherry-red seats, and the ride across the moss-colored Hudson River. But that day, he was furious: a job that should have been simple had ruined his week. Herbert was then the co-owner of a printing company named Pantone, and he was fuming because when he sent out orders for ink, he couldn't predict what color he would get.

Back then, "every designer had about a half a dozen color books in his drawer. Every ink company used different sets of pigments which react differently under different lights," he told me, speaking from Palm Beach, where he now lives. Because of the confusion, he often had to mix his inks by hand. Why, he wondered, couldn't everyone just use the same color book?

With that question, he began to imagine a kind of alternative reality, where printers and ink makers and ad designers all spoke a universal language—a simple idea that would change the publishing industry. Each shade of pink or purple, he realized, could be expressed as a number. Under his scheme, "if somebody in New York wanted something printed in Tokyo, they would just simply open up the book and say, 'Give me Pantone 123,'" he said. That way, you could be sure that 123 (a daffodil yellow) would look exactly the same the world over.

Herbert created a sample page to show how the system worked and sent it to ink makers. Fifty years later, he still owns a copy of that page of orange squares, each one a slightly different shade.

It wasn't easy to make Pantone colors the industry standard. Herbert had to operate like a politician, schmoozing, trading favors, and building up alliances with graphic designers and ink companies. Finally, he took out an ad proclaiming that the Pantone Matching System would debut in September of 1963. And because the system made so much more sense than the chaos that preceded it, Herbert's scheme did indeed become that universal language of color.

By the 1970s, Pantone was reaping more than a million dollars a year in licensing fees from its copyrighted system. Pantone was also one of the first companies to use data mining to predict trends—once the colors had become numbers, it was far easier to track demand and monitor markets. "We had a consultant who would get a committee together and find out, say, 'What colors are showing up in Milan? What colors are showing up in Paris?' It seems that a lot of designers all decide that coffee brown might be a good color in the same year," Herbert told me, and the company had to continually add colors to reflect evolving consumer tastes.

As the Pantone system spread from the advertising world to textiles to food science, it was put to some unexpected uses—like defining the color of a Ben and Jerry's brownie. "I have matched color charts for wine," Herbert said. "I matched color charts for anemia blood samples, and for walnuts and strawberries and goldfish."

The Pantone Matching System continues to generate millions of dollars a year and has worked itself deeply into the way designers think. (Calvin Klein reportedly kept a Pantone swatch in the kitchen to show his chef what color he wanted his coffee to be.) What's most amazing about Herbert's invention is that he seems to have created so much out of so little; the earliest version of his invention was just one page.

Herbert had been able to see the value hidden in what appeared to be *nothing*. Color books were free back then. The ink makers sent them to you, and you threw them into the back of the print shop. So Herbert had to imagine a possibility that would have seemed crazy at that time—that people would *pay* (a lot) to buy into a language of color. To invent, he had to be able to un-think the world as it was. He had to recognize that the normal state of things, the boring and unremarkable, actually contained the seeds of an empire.

Herbert's story is a lesson to us all: to transform the world, you can start small, with whatever is sitting on your desk right now.

One morning, a brain scientist named Alice Flaherty picked up her toothbrush and stared at it in wonder. "I brush my teeth *every single day*." The thought thundered into her mind like a revelation. Alice had spent years treating people with neurological disorders, but now she was experiencing the same kinds of symptoms that haunted her patients. (I use her first name here because—full disclosure—she is a longtime friend of mine.) Even the most ordinary moments—like squeezing toothpaste onto a brush—could inflame her imagination, convince her that she'd made a major creative discovery, and compel her to record her "revelation" in pages of notes.

A tragedy had launched her into a temporary state of psychosis; she had given birth prematurely to twin boys who died, and in a spasm of grief, she began covering paper and Post-its and walls with handwriting, blotting out blank space with words, staying up for days, sick with language. She was possessed by a condition called "hypergraphia," a form of compulsive writing that accompanies a psychotic break. She eventually recovered, but a few years later, she gave birth to healthy twin daughters—and even though that pregnancy turned out to be a happy one and her babies were healthy, it triggered another bout of mania. Alice told me that during that time, she suffered from an "excess of meaning." She saw so many possibilities—and was so excited by them—that she couldn't function.

Though she recovered and is now chief of the brain-stimulation center at Massachusetts General Hospital, she describes herself as still distracted by enthusiasms and revelations. You merely have to name a topic, from bike helmets to bedbugs, and she's off, making observations about cultural biases, asking questions, citing Shostakovich and Danielle Steel. A slim woman in delicate glasses, she studies you urgently as she showers you with compliments, so that somehow all the ideas she's stringing together seem to be yours.

In her office, she keeps a wooden box with a sliced brain under her desk—she pulled it out of the trash after another doctor threw it away. Her shelves are packed with curiosities, reminding me of those private museums that aristocrats assembled in the seventeenth century when the study of nature was accomplished by collecting oddities. Her cabinet of wonders includes a preserved garfish, a snapping turtle skeleton she found in the woods, and a "low-carb" communion wafer that she bought off the Web. When she talks about her patients, she does so with similar enthusiasm, as if she has assembled a human cabinet of wonders. She pours out sympathetic case histories of "the itchy guy," "the ex-con," and "the patient I accidentally made a compulsive gambler."

In 2014, Alice had become mesmerized by her own invention, which she cobbled together in her kitchen out of metal parts she lathed in a friend's workshop, as well as a bicycle wheel and a treadle pump. "I decided that I needed to build a machine that unravels sweaters. You can buy a cashmere sweater at Goodwill for $4.99, and then unravel it and you end up with $100 worth of cashmere yarn," she said. She spent hours in a friend's workshop tinkering with her foot-powered yarn-recycling machine. "For some reason I got obsessed with this. It was a glitch in my head. I was intoxicated by the idea that I could turn this cheap sweater into an expensive commodity." She felt magical, like Rumpelstiltskin spinning straw into gold.

As a psychology researcher, Alice studies these "glitches" in our heads that lead to inventive breakthroughs. She is particularly fascinated by the ways in which brilliant inventions resemble crazy delusions. "The production of ideas or actions that are novel but useless . . . is part of the vernacular conception of madness," she wrote in one scientific article. She points out that the word—*useless*—often is what divides the sane from the insane. If you spend years working on a weird machine that you can't sell to anyone, you might be written off as a lunatic, a kook, or a failed folk artist. But if you make millions of dollars, you're hailed as an innovator.

Alice likes to point out that Chester Carlson, the inventor of the technology behind the now-ubiquitous copy machine, came very close to being labeled a crank. In the 1930s, he had worked in the patent office of an electronics firm and "I frequently had need for copies of patent specifications and drawings, and there was no really convenient way of getting them at that time," Carlson wrote later. That led to a revelation that struck his contemporaries as mad: Carlson believed that thousands of office workers could be replaced by a machine, and that—in effect—a fortune was hidden in the acres of secretaries that stretched all across America. Carlson built a photocopier, and then he spent years trying to peddle his machine. "How difficult it was to convince anyone that my tiny plates and rough image held the key to a tremendous new industry. The years went by without a serious nibble. . . . I became discouraged and several times decided to drop the idea completely. But each time I returned to try again. I was thoroughly convinced that the invention was too promising to be dormant," he wrote. Of course, his invention would eventually kick off the Xerox empire. But in a slightly different world, Carlson might have remained a crackpot with a useless machine. This gets us back to Alice's insight about the nature of the most significant inventions: they ask all of us to embrace a *slightly different kind of sanity*, one in which the new technology makes sense. The something-out-of-nothing inventions can be the most difficult to disseminate because you must convince thousands or millions of other people to redefine *useless* and see value where none existed before.

At the same time, the something-out-of-nothing impulse lies behind some of the most resilient and resourceful businesses, the kinds that flourish even in the midst of economic disaster.

Everyday MacGyvers

In the late 1990s and early 2000s, two academic researchers—Ted Baker and Reed E. Nelson—decided to investigate a mystery. They had noticed that some local businesses seemed to be especially resilient and were able to flourish even during economic downturns. Why?

To find out, Baker and Nelson undertook fieldwork in the coal-mining country of southern Illinois, a region hard-hit by unemployment. There, amid the dead streets and shuttered shops, they discovered dozens of wizards who tinkered, recycled, and invented. Unusually ingenious, these local entrepreneurs continually found new ways to turn worthless materials—like trash, spare parts, polluted land, and even fumes spewing out of sinkholes—into stuff that other people needed.

For instance, Tim Grayson owned farmland that was pocked and polluted with the remnants of old coal-mine tunnels that leaked methane. What other farmers might have regarded as a problem, Grayson turned into a solution. He and his partner drilled into the mine shafts and then channeled the gas to a generator that had been rejiggered to burn methane. The work had been dangerous—Grayson had survived several small explosions—but it provided him with free electricity and heat. He capitalized on this by building a greenhouse for hydroponic tomatoes and then a fish farm.

The researchers noted that Grayson's success rested on his ability to redefine what other people called a "dangerous nuisance" into a resource; the tunnels became his own private power plant.

Meanwhile, another local fix-it man named Jim Roscoe had found his own way of exploiting the abandoned mines in that region. He learned that local power companies had to send their technicians to check on the high-voltage electrical lines that crisscrossed the mucky underworld. That inspired him (and a friend) to invent a tool to troubleshoot the underground lines, and he sold a few of them to the power companies. At the same time, he continued to ply his skills as a plumber, TV repairman, Sheetrock hanger, carpenter, and musician.

The researchers also studied Jim Jarvis, owner of a motorcycle repair shop. Every day, a few hours before the shop closed, Cycle Tech functioned as an informal community center and bar. People would show up with beer and stay to tinker with their bikes, gossip, and trade parts. Customers would even pick up wrenches and help with the shop's repair jobs. The researchers observed that people like Jarvis had a way of creating an ecosystem where one person's trash became part of someone else's bike, and where customers became collaborators.

In other words, the dozens of resilient businesses in the study made their living off of the stuff that most people regarded as useless—or the resources that other people simply didn't know existed (like methane in abandoned mines)—and by tapping the talents of their customers and friends. Many of them maintained scrap heaps that they mined for parts, preferring to invent their way out of problems rather than buy prefabricated solutions from suppliers. And rather than specializing, most of them were engaged in a never-ending discovery process. Hunter-gatherers in an industrial wasteland, they were always looking for new opportunities in junkyards and sinkholes.

When your experiments are free, and involve trash, you are emboldened to take risks. This is the lesson we can learn from the renowned Building 20.

Building 20

In 1991, Tim Anderson wandered onto the MIT campus and discovered the trash of his dreams. In a hallway, he spotted an oscilloscope that must have cost a fortune; it had a note taped to it that said FREE. PLEASE TAKE. Tim was a connoisseur of techno-garbage, but he'd never seen anything like this before. As he began foraging around the MIT campus, he discovered even more treasures: old Teletype machines, castoff robots, and copper wire. So he moved onto the MIT campus to harvest its garbage, living for years as a squatter there.

In the mid-1990s—when I met him—Tim slept on a blanket tucked underneath a robot in a back room in the school's legendary Building 20. I don't think anyone in our circle of friends, least of all Tim himself, would have guessed that one of his whimsical projects would eventually help to launch a billion-dollar industry. With his ruff of blond hair and all-season sandals, Tim could pass for a surfer who had gotten lost on the way to the beach. He had never actually attended MIT as a student. Instead, he had learned sword making in Japan, had sewn his own wetsuit, and had worked in a harness shop repairing the gear that dairy-farm cows wear on their udders. Tim listed that gig as "cow-bra repair" on his résumé—which, he said, helped to keep him happily unemployed.

And so he'd landed in Building 20, the university's unofficial junkyard and skunkworks. Slapped up as a temporary structure during World War II, it had once housed the radiation lab; but now the building was slated for demolition (it would be torn down in 1998) and no one minded what you did inside it. You could tear a hole in a wall or skateboard down the stairs, or bring your sleeping bag and move into a spare room.

"One of the rooms [in Building 20] was an anarchist utopia where you could just show up and tinker," Tim told me. "The room was so full of junk that you could only open the door partway, so you had to squeeze yourself through it to get in." Officially, this junkyard room belonged to a student club, called MITERS. Unofficially, this was Tim's apartment.

Building 20 happens to also have been one of the most prolific hubs of invention in the twentieth century; in its heyday, it was the site of revelations about microwaves, electronics, neurobiology, and even linguistics. (Noam Chomsky once worked there.) It had nurtured at least nine Nobel Prize winners. You could argue that it was also the birthplace of a new idea about invention—the belief that everyone should have access to public R&D labs.

So why did the grungiest building at MIT become one of its most creative hubs? The answer to that question tells us something essential about the nature of creativity.

One day in 1992, when Tim was foraging on the MIT campus, he plunged into a Dumpster behind Building 35. When he came up for air, he met Fred Cote, who worked in a lab nearby, and Fred offered to show him around. There, Tim had his first glance of an entirely new kind of technology.

Tim had never heard of 3-D printing. Back then, in 1992, fewer than five hundred 3-D printers were in use around the world. They cost up to half a million dollars and could produce only a limited repertoire of objects made out of a special kind of plastic. The researchers in Building 35 wanted to revolutionize these machines to make them more versatile. And they needed an electrical engineer.

Tim hadn't even *wanted* a job. But when the guys in Building 35 saw a robot he'd made out of trash, they talked him into working on their project and introduced him to the leader of the lab, Ely Sachs. Soon Tim—despite his lack of credentials—ended up crawling around inside Sachs's machines. He realized that he had just stumbled on the most challenging and fascinating problem of his lifetime—one that had stumped the top engineers and scientists in R&D labs around the world.

Sachs was trying to perfect a method of 3-D printing that involved a fine powder that would clump together to form the object. The method promised to open up many possibilities in the field. "For instance, how could you print a design on the inside of a solid sphere? You couldn't do that back then. Not with any machine," Tim explained. The goal was to allow people to print objects out of new kinds of materials (like ceramics) and in new shapes.

But this precise 3-D printer, based on Sachs's idea, was maddeningly hard to build. "We were spraying this liquid, tiny droplets, between two high-voltage plates. And I spent the next three years basically inside that big machine wiring up a whole lot of electronics," Tim said. "This was an entirely new process that no one had ever tried before." At that time, the finicky machine had to be nursed by a team of people who would keep it running.

Along the way, Tim and one of his friends in the lab—a grad student named Jim Bredt—decided to launch their own experiments in the trash heap of Building 20. "In Ely's lab, we'd been trying to build a machine that could print super-high-quality parts," Tim said, because corporations needed 3-D printers to speed up the manufacture of engines and aeronautics devices. "But Jim and I wondered what would happen if you didn't have to worry about perfection. We wanted to make a fun machine."

Specifically, they imagined creating 3-D printers as "party spectacles" that they could use to amaze friends. For instance, what if they could invent a machine that would print out a tiny version of someone's head—a Shrunken Head Machine? Wouldn't that be cool?

Tim had realized that he could repurpose parts from ink-jet printers and use those as the core of their fun machine. Instead of starting from scratch, he and Jim would use old technology in new and unorthodox ways. "To me, that idea opened up a sense of possibility," remembers Jim. "And that's how we built the Kitty Litter Machine."

Made from the guts of old printers, the machine slung Elmer's Glue onto kitty litter, so that the green powder fused together into objects. "We scanned the face of my then-girlfriend and tried to print out her head in kitty litter," Tim told me. "We wanted it to look like a portrait but it ended up looking like a big turd." Their 3-D printing experiments seemed like "a dismal failure," according to Jim.

In the winter of 1994, Tim began working on the Candy Machine. He started with a butchered HP ink-jet printer and spent hours figuring out how to fool the computer chips inside the machine, which were programmed to shut down the ink-jet mechanism when the printer had run out of paper. Then Jim and Tim began thinking about what kinds of materials would work best for creating objects. They performed their research in a Purity Supreme grocery store, where they grabbed powdery foods that they hoped would clump together to form a three-dimensional shape. Soon, Tim was shooting water through a printhead at a pile of sugar. As soon as one layer of the sugar dried, he did it again, printing in exactly the same spot. Then again, and again. In this way, he built up the letters layer by layer until they formed tiny towers. When the sugar hardened, he had produced 3-D letters that spelled out MIT. (As Tim liked to point out, this was also a backward spelling of his own name.) "I took an X-ACTO knife and popped the letters off the cardboard and then I had these little objects. I thought, 'Oh, wow. Incredible,'" Tim said.

A few days later, Jim and Tim discovered that Sweet'N Low worked better than sugar for 3-D printing. "We were going around restaurants stealing Sweet'N Low packets off of tables, and then dumping it out of the pink packets. We absolutely did not want to have any poisonous stuff in our printer. And it turned out that Sweet'N Low would stick together in really precise shapes," Tim said.

Fortune 500 companies were spending millions of dollars to develop 3-D printing technology; meanwhile, Jim and Tim had just invented a machine that could fabricate an object in fifteen minutes—a speed that seemed impossible back then. And they'd spent less than $100.

Soon a series of would-be partners made the pilgrimage to Building 20, threading their way past piles of robo-junk, banana peels, and dirty laundry to watch them demonstrate their 3-D printer.

"We'd turn on the printer," Jim recalled, "and it would make noises for a few minutes and then it would stop."

"Is it broken?" the investor would ask.

"No, it's done," Jim or Tim would say, and then reach into the printer and pull out an engine part that the printer had molded out of a pile of Sweet'N Low.

Jim and Tim signed a deal with two business partners and founded Z Corporation to develop this technology into a product. In 1997, Z Corp released the world's fastest 3-D printer. The company became a leader in the nascent industry.

Tim and Jim's story points to the quixotic nature of creativity. "One of Tim's major innovations was to realize that a 3-D printer was just a more-complicated version of an ink-jet printer," Jim said. "That was the idea that inspired us to butcher ink-jet printers to find out what was inside them." And once he and Jim began playing with castoff Teletype machines, they felt free to pour ketchup or sugar or flour into the printhead—because when you're tinkering with trash, you can take risks.

When I called Jim Bredt to arrange an interview with him for this book, he suggested that we meet at a hackerspace in Somerville, Massachusetts, called Artisan's Asylum that has become his second home. Jim now holds down a day job as chief technology officer at Viridis3D, and at night he continues to tinker with trash. When I found him at his workspace at Artisan's Asylum, he was sitting behind a dressmaker's mannequin that he had set up to display a polystyrene gown covered in rubbery replicas of human teeth. This was one of Jim's side projects: he'd found some old dental molds that had inspired him to make this unusual piece of couture.

There must be some magic in his creative method. A 2014 report on the 3-D-printing sector cited Jim Bredt as the most-patented inventor in the industry. In other words, don't dismiss the toothy dress; it might inspire some new kind of technology, or Jim's next printing invention.

It's this same kind of ingenuity that can enable us to outsmart hunger and disease.

The Power of Nothing

In 2013, Bill Gates confessed in *Wired* magazine that he has become obsessed with fertilizer. "It's the kind of topic I have to remind myself not to talk about too much at cocktail parties, since most people don't find it as interesting as I do," Gates wrote. "A full 40 percent of Earth's population is alive today because, in 1909, a German chemist named Fritz Haber figured out how to make synthetic ammonia." That ammonia could, in turn, be used to create synthetic fertilizers that gave farmers a way to turbocharge the growth of their crops. A few years later, the German chemical engineer Carl Bosch developed a method that allowed factories to mass-produce those synthetic fertilizers. The Haber-Bosch process has been named the most significant invention in history, if you measure it by lives saved; during the population explosion of the twentieth century, it helped farmers feed millions of people who otherwise would have died in famines, and we now depend on the agricultural bounty it produces to sustain us. It may seem paradoxical that Gates—an icon of the digital revolution—would become fascinated with soil, but the Haber-Bosch process, he wrote, offers a lesson about "where you can make the greatest difference, where every dollar you spend is liable to have the greatest impact."

The Haber-Bosch process shows how humans can deploy invention to overcome the biggest problems that threaten our survival. If we can transform air into bread, what else might we be able to do?

Again and again, people have found ways to transform "nothing" into a new resource, discovering that the trash or filth contained a secret of immense value.

In 1865, Jules Verne wrote a story about a spaceship made out of aluminum—an idea that at the time sounded as whimsical as constructing a moon colony out of solid gold. Aluminum, derived from an abundant ore, was then so difficult to process that only tiny lumps could be produced; according to legend, Napoleon III ate off of aluminum plates because they were more valuable than gold. But then a series of inventors made breakthroughs in the processing of the metal, and soon aluminum became plentiful and cheap. Just as Verne had predicted, this miracle substance turned out to be ideal for flying machines; the boom in aluminum helped to spawn the aeronautics industry.

Meanwhile, Carl Djerassi, an inventor of the birth-control pill, harvested hormones from Mexican yams; the lowly tuber also contributed to our well-being as a species, by allowing us to fine-tune the timing of pregnancies.

Most recently, researchers have found treasure in the most unlikely place: feces. New genomic tools allow researchers to catalog and study the microbes that live in our guts, and this ability to explore our internal ecosystem has led to some of the most illuminating breakthroughs of the past few years—and potentially a new trove of medicines. In fact, one of those new medicines is already here: people who are suffering from a gut infection called *Clostridium difficile* can be cured, often within a matter of hours, with an infusion of feces from a healthy donor. We have turned excrement into medicine—and that's a remarkable something-out-of-nothing coup.

Over and over again, people have reshaped the environment and manufactured what we need out of whatever we have. The results are often far from perfect; for instance, the fertilizers produced by the Haber-Bosch method ushered in factory farms and a system that depletes the soil of its microbial diversity. Still, reviewing some of the great breakthroughs of the past, you can only be awed by our ability to use whatever is plentiful. To do that requires more than scientific discovery; it also depends upon one of our most unusual gifts as a species. We can imagine the future, time-traveling inside our own minds and addressing problems that don't yet exist. And so, let's teleport into the future of this book, Part III.

PART III

PROPHECY

8

THE PONG EFFECT

THE MODERN CELL PHONE IS A CONFECTION OF TECHNOLOGIES, all rolled into one irresistible bonbon. It depends on tiny computer chips and towers strung with chandeliers of antennas, and it arose from breakthroughs in a variety of fields, from telephony to touch-screen engineering. So it would be wrong to credit any one person or one moment as the origin of the cell phone we use today.

And yet, the *experience* of using a hand-held communicator was also its own kind of invention, and that experience began to emerge in a specific time and place. In this chapter, we will follow an engineer named Martin Cooper (first mentioned in the introduction of this book) who developed a new idea about how people could communicate. Back when the telephone was still bolted onto the wall and tethered by wires, Cooper liked to joke that one day everyone would be issued a phone number at birth that would be retired at death. That idea seemed wildly improbable in the 1960s, but he stuck to it and devised strategies to make it come true. "I have a mantra that people are naturally, fundamentally, and inherently mobile," Cooper told me, and he realized that this "innate" human desire would drive the evolution of the phone.

When you invent, you have to think about the future; you're making a bet about what people *will want* ten years from now, or what they *would want* if it were possible to overcome certain limitations of the present. In Part III, therefore, we will explore invention as it relates to prediction, prophecy, and science-fiction dreams. What sorts of people first imagine that future? How do they do it? And how do they communicate their vision to the rest of us? We'll meet a series of Inventologists who searched for the philosopher's stone: a formula that would reveal how technology evolves so that it's possible to jump ahead of the curve.

As for Martin Cooper, he has come to believe that it's most important to design a new *experience* that just about everyone will crave. But here's the tricky part: before that experience exists, few people understand it. They may even hate the idea. So you have to put the machine in their hands in order to awaken a new desire and explore new possibilities. That's just what happened at Mount Sinai Hospital in New York City.

In the 1950s, as you walked through the halls of a hospital, you were assaulted by the noise. Loudspeakers hung in the hallways and operating rooms, broadcasting a steady stream of announcements to the nurses and doctors who rushed below—"Dr. Harrison, please report to the ICU. Code Blue, CODE BLUE."

It was frightening for the patients to hear about emergencies, so a team of engineers from Motorola came up with a proposal: What if the nurses and doctors carried small radio devices, similar to walkie-talkies, that let them speak to one another as they moved around the wards? The engineers decided to run their experiment at Mount Sinai Hospital, which was then a congregation of buildings and a nursing school that sprawled over several blocks.

Motorola's experiment—which began in 1955—was a vast undertaking; it required miles of antenna wire and a transmitter that beamed the messages to the hand-held pagers that were the size of cigarette boxes. But the results were miraculous. When a central dispatcher placed a call, a surgeon would reach into a pocket, listen to instructions, and rush to exactly the room where he or she was needed, prepared to handle whatever emergency was unfolding; meanwhile, a nurse could show up at a bedside in seconds with just the right medication; and a new mother could use the pager to listen to her child in another room. It was as if the hospital had been given cell phones decades ahead of everyone else.

Of course, those early communicators *weren't* phones; once the staff left Mount Sinai and walked out onto the street, their radios would crackle into uselessness. Outside the range of the signal, the handset could not send or receive messages. Back in the 1950s and early 1960s, if you wanted to move around the city with your radio-communicator, you needed a car or truck outfitted with a suitcase-size box that contained the transmitting equipment. That's why—despite the experiment in Mount Sinai—it was hard to imagine that mobile phones would ever become tiny devices worn on the body. On the Sunday funny page, Dick Tracy spoke into a wristwatch radio, but that technology seemed as remote as the Atomic Light Unit he threw at bank robbers. The general assumption back then was that mobile phones (when they arrived) would have to be built into vehicles, because the underlying technology would take up so much space.

But inside the halls of Mount Sinai, a rift in time had opened up, and the staff began to live in the future. Now that everyone was using hand-held communicators, the technology seemed obvious and inevitable. The medical teams began finding their own ingenious ways to use the pagers; for instance, a surgeon could hide out in a back room to catch a few hours of sleep knowing he or she would be woken up by the communicator. In Chapter 1, we saw that the early adopters of a technology often experience a condition that might be called Martian jet lag; they're the first to suffer from the problems associated with the technology, and their insights—born out of frustration—often inspire spinoff inventions. In this case, the doctors and nurses at Mount Sinai devised clever new ways to take advantage of their radio system even while they struggled with the science-fiction problems it created. The hand-held radios were unreliable—you never knew when the transistors would burn out and your communicator would go dead. Since the medical staff now used the Motorola radios to choreograph their movements around the hospital, these breakdowns caused a futuristic distress. Today, of course, most of us know what it's like to panic about a dead phone; it's a peculiarly twenty-first-century freakout: when the screen goes dark, you suddenly realize how helpless you are without it. But in the 1950s and early 1960s (when Motorola tested its system in the hospital), people were just beginning to wake up to the potential—and the pitfalls—of modern communication. "Doctors can be real prima donnas," Martin Cooper said, describing how they threw tantrums when their radios broke. One doctor even hurled his against the wall.

For Cooper, who was a senior engineer and then a product manager at Motorola, the experiment at Mount Sinai opened his eyes to how addictive hand-held, wireless communication could be. One day, a delegation of Motorola engineers told the doctors, "Look, we made a mistake, we were a little premature, let us take the system back. We'll improve it, and when it's ready to go we'll come back and give you a better system," according to Cooper. But the doctors refused to hand over their devices. "They now could not conduct the business of the hospital without pagers," according to Cooper. This experience gave him his first glimpse of the future of telecommunications: once you begin wearing a communicator on your body, you change your habits, and soon you can't function without it. If you "try to take your product away from somebody [after they've started to use it], and they refuse to give it up, then you know that you were successful," according to Cooper.

His second inkling came in the late 1960s. At that time, Cooper and his colleagues at Motorola began testing radio communicators built for workers at airports; these devices, bigger than the models used in hospitals, weighed several pounds. In theory, the communicators were too heavy to carry around in your hand all day, so Motorola manufactured elaborate holsters so that the gizmos could be stowed on the workers' backs. "We'd gone through a lot of trouble to make holsters, devices that you could carry like a backpack," but it turned out that the airport workers rarely put their communicators away, Cooper told me. These workers were not exactly Lead Users, because as far as we know they didn't invent anything. And yet, they did figure out how to do their jobs more efficiently by engaging in a constant stream of conversation with their managers and coworkers. As he observed them, Cooper became even more convinced that the future of phones would be in our hands.

Cooper was lucky, because his job put him in a position to watch how people behaved in these environments that simulated the future; thus, he was able to undertake the kind of ethnography and fieldwork that are crucial to understanding the minds of users (as we saw in Part I). The Motorola experiments gave him a chance to study truck drivers, surgeons, airport workers, and real-estate agents as they interacted with new communications devices. "Most of the invention doesn't happen in the mind of the inventor; it happens in the observation of how people use . . . the product," he said. He himself had been wearing a pager long before they were available to the public, so he was already living in that future where he could communicate from anywhere.

But even if the hand-held phone seemed inevitable to Cooper, there were enormous hurdles to overcome. In the 1960s, America's phone system was controlled by the AT&T monopoly, and it looked as if the Federal Communications Commission (FCC) might hand over control of key portions of the cellular spectrum to Ma Bell.

AT&T engineers had spent decades planning out a future in which cars would become mobile phone booths. In the 1940s, the engineers at AT&T's Bell Labs had designed a system called "cellular telephony" that involved splitting up each city into small "radio stations" that would communicate with each other. In the 1960s, AT&T revived this idea; its vision of the future can be glimpsed in a patent written by inventor Amos Joel: his diagram of this system showed radio towers beaming signals to individual cars equipped with powerful antennas. This vehicle-based scheme made a lot of sense back when the cell phone units were so bulky that you had to carry them around in a suitcase.

In the early 1970s, the FCC gave companies a chance to lobby for a piece of the action, and it opened up some of the radio spectrum to encourage the upstarts to demonstrate their cellular devices. Cooper and his colleagues watched these developments with growing desperation. It seemed likely that AT&T would win the scramble for the cellular market—Ma Bell was still closely tied to the government. "Around November of 1972 we got word that the FCC was about to make a decision," Cooper remembered later. "The worst possible thing that could happen to us was for AT&T to take over our business. So we decided in November that we were going to go to Washington" to appeal to the FCC committee members.

That's when a plan began to take shape in Cooper's mind. "We had to do something dazzling to get the attention of people," he told me. Simply describing the phone wouldn't be enough because "that's almost like science fiction." Back then, the future of the cell phone had been scripted by AT&T with its vision of car-based cell phones. To prove that a hand-held phone would be possible, Motorola needed to change the conversation, and a mind-blowing demo could do that. "It's not an invention if you can't actually make it," he said. But "if you [could] put a device into somebody's hand and let them walk around the room, talking on a [phone] without a wire on it, that would capture . . . attention. That was the genesis of that idea of the first hand-held cell phone."

Cooper suggested to his bosses that the company throw itself into building the demo, and after serious deliberation the managers decided to go ahead. Because of the advent of miniaturized transistors and smaller batteries in the early 1970s, Cooper thought it might finally be possible to build a hand-held wireless phone that could call anyone in the world.

Motorola was betting everything on that demo. "We literally shut down all engineering at Motorola. There were literally hundreds of people working on this system" in order to build the first hand-held cell phone, Cooper said. Credit should therefore go to the top management for taking what must have seemed like a perilous risk. Cooper said that the CEO, Bob Galvin, "was betting the company" on the cell phone.

The demo model would be starring in newspaper photos and press events. It had to look as if it had just been beamed in from the year 2000. So Cooper asked Rudy Krolopp, Motorola's industrial design director, to create a mockup of a hand-held phone.

"What's that?" Krolopp replied, puzzled by the very notion.

But two weeks later Krolopp's team presented Cooper with several designs. The pretend phones—all of them white and sleek—seemed to have emerged from an episode of *The Jetsons*, with sliding panels, flip-up lids, and cartoony antennas.

Cooper took one of the models to his team of engineers and said, "Could you make this into a real phone?" It turned out they could—but the "phone grew and grew. By the time we got it done, it was what they called the 'brick.'" Still, it worked.

On April 3, 1973, Motorola called a press conference at the New York Hilton on Sixth Avenue. Before he went inside to give his presentation to the journalists, Cooper paced along the sidewalk, fiddling with an object that looked something like a child's boot, with pushbuttons and an antenna. A crowd gathered around him, gawking, as he demonstrated how to make a call from the sidewalk.

After this bit of street theater, Cooper met with journalists inside the Hilton. The first cell phone had all the sex appeal of a doorstop.

Still, it was a triumph of engineering. To prove that the phone wasn't an elaborate fake, he handed it around. One reporter called Australia and was astonished when her mother's voice came out of the device. The Motorola engineers had created a simulation of the future by cobbling together technologies of their own time; behind the scenes, the workings of the cell phone were very much patched together. The hand unit sent radio signals to a base station in a building across the street; from there, it connected to AT&T's landline network. There was still no national network of cell towers, still no infrastructure to support this technology. On top of that, the phone conked out after twenty minutes of battery life and was so heavy that you'd want to hang up long before that. And yet, it worked well enough to prove Cooper's point: this phone could exist.

The demo captured the attention of the press, and Cooper believes that it helped keep the FCC from handing over all of the radio spectrum available for cellular technology to AT&T. The federal agency finally opened up the radio spectrum in the early 1970s. When it allowed several companies to compete in the early 1980s, Motorola jumped in to sell the first-ever commercially available hand-held mobile phone, the DynaTAC 8000X. The original "brick phone" soon became a film star and symbol of yuppie excess. In the 1987 film *Wall Street*, the character Gordon Gekko strolls on the beach at sunrise, snarling into his brick. "This is your wake-up call, pal," he says. Soon enough, everyone else would get the wake-up call, too. Within a decade, the hand-held phone would conquer the world.

Of course, we should also give credit to Chester Gould, the creator of Dick Tracy, who inspired Martin Cooper and his generation to imagine a wristwatch-like communicator; in the 1950s and 1960s, kids could even wear the toy Dick Tracy gadgets equipped with buzzers and Morse code clickers. Gene Roddenberry and the set designers of the original *Star Trek* also shaped fantasies about the phone; in the

1960s, the crew of the *Enterprise* were already carrying flip-top communicators. I still remember the cool way that Captain Kirk would open up his phone with a flick of the wrist that was reminiscent of a smoker tipping a cigarette out of the pack; it made the communicator seem oddly illicit and enviable. In other words, the science-fiction fantasists of the mid-twentieth century gave people their first taste of a wearable communication device—and that widened their sense of possibility. (As we will see in the next chapter, stories and movies and cartoons are often the most important sources of technological vision.) Media-bait demos like Cooper's and early products like the Motorola DynaTAC further awakened those desires. The origin of any radical and ambitious technology is almost always double-stranded. It often begins with wires and blueprints, but also with visionaries who transform our expectations.

Cooper told me that he taught himself to explore imaginary worlds and futuristic possibilities. As a child, "I read a lot of fantasy and mythology, and when I got older I started to read science fiction. So my mind has been floating around forever, which is not a very good attribute for somebody who wants to be an executive," he said. But that powerful mind's eye was Cooper's head lamp, one that allowed him to see into the future. "When you combine the attribute of dreaming with an interest in science and how things work (and that's always been important to me), you're just naturally constantly thinking of how things could be done differently," he said.

Pong-i-ness

Cooper believes that certain kinds of machines are so much fun, so addictive, and so perfectly suited to the human brain that they create entirely new tastes that shape a generation. When these new technologies first appear, they usually look ridiculous (like the brick phone) and it's easy to laugh them off—but even as they're taking baby steps, they transform our expectations and win converts, paving the way for an entirely different relationship with machines.

"We shape our tools, and thereafter they shape us," the media scholar John Culkin once observed. The most ingenious technologies teach us to form new habits. And once you've acquired that habit, the technology is a part of you, and you can't imagine living without it. Cooper calls this phenomenon the Pong Effect.

If you're a Gen X-er, you may have played Pong in shag-carpeted basements in the 1970s, twirling the dials on the game console as the "tennis ball"—a white square—drifted across the TV's black screen. Crude as it was, Pong blew our minds as kids. Why? Because we'd been accustomed to passively watching whatever was beamed into the TV. But now, suddenly, you could reach inside and make things happen on the screen. Pong was a kind of gateway drug—it gave us a taste of what would be possible when more sophisticated video games arrived in our basements. In Tim Wu's book *The Master Switch*, he—like Cooper—points to Pong as that rare kind of technological confection that created an appetite for more products like it, opening up an entirely new industry. "Looking at today's PlayStation, who would guess that Pong had once been a transfixing game?" Wu wrote. "Or, for that matter, that in the age of hi-def, YouTube, with poorer resolution than television of the 1940s, could have caught on as it has? In fact, the primitive prototype is typical in the founding stage of a new industry."

We are accustomed to thinking about large R&D labs as the birthplace of the radical technologies that shape our habits. The cell phone, for instance, required decades of engineering, billions of dollars, and a small army of lawyers and legislators in order for it to burst into existence. And yet, not every taste-making technology has been born in a Bell Labs or a Motorola. Pong is proof of that. It was developed by a scrappy duo—Nolan Bushnell and Allan Alcorn—who tested it out on a cheap black-and-white TV set. Like Cooper, they were designing an *experience* rather than just a technology.

The Pong Stage suggests that technologies don't need to be slick or fully realized to catch on. Instead, they must open up a crack in the world, like Alice's looking glass opening a door into Wonderland; we pass through these machines as if they were portals to the future. The Pong-iest machines are predictions made out of wire, glass, and molded plastic. They let us taste, feel, and smell what could be, which is a sort of sorcery.

But to pull off this voodoo, inventors often have to engage in a lot of hardheaded technological forecasting. Cooper and the engineers at Motorola, for instance, had to *predict* the evolution of transistors and batteries, so that they could leap ahead and take advantage of opportunities that would open up in a few years. How are some people able to ignore the limitations of their own time and anticipate what will be possible? In the next chapter, we will dig into that question.

9

THE WAYNE GRETZKY GAME

IN 1933, A CAMPUS NEWSPAPER PUBLISHED A BIZARRE ESSAY that was "written by some future vice-president of MIT." Perhaps the strangest thing about this stunt: it was dreamed up by the actual vice president of MIT, Vannevar Bush, a buttoned-up engineer in a tweed waistcoat, perpetually shrouded in a cloud of pipe smoke. He looked less like an inhabitant of the future than like a throwback to the Edwardian era. And yet, Bush used his essay to mock the primitive machines of his own time, scoffing at the way people in the 1930s suffered the "incessant clatter of typewriters" and lived in houses that were anchored in one place and couldn't move—how silly! He held particular scorn for the libraries of his own time, which he criticized as more like museums; you wandered through them lost in a maze of shelves. To find the information you wanted, you had to "paw over cards, thumb pages and delve by the hour."

Around the time he wrote that essay, Bush's mind was full of libraries; he was frustrated that this most important tool of human thinking should be so primitive. But he also brimmed with hope, because a dazzling new technology was in the offing: microfilm! Today, of course, the microfilm machine is a relic; with its gears and levers, and the hand crank that turns the reel of film, it is the Tin Lizzie of information science. But in it, Bush detected the potential for a revolution. Microfilm could miniaturize the world's knowledge, and once libraries became small, they could become ubiquitous; ordinary people could own thousands of books, stored in a desk drawer. He set to work engineering a "selector" that would allow people to zero in on the information they desired; in retrospect, it seems like an attempt to fabricate Google out of wood, steel, and reels of film. Unfortunately in 1939—the tail end of the decadelong Great Depression—he couldn't hope to fund this ambitious project. Though Bush was now rising to a position of great political influence (he would direct America's R&D effort during World War II), he still had to "scramble for money to support MIT, his department and his students," according to historian Colin Burke.

And so, lacking the resources or time to construct an actual machine, Bush retreated into his imagination and challenged himself to predict how the personal library would work once it arrived. In an unpublished essay in 1939, he described the Memex, a futuristic machine that would allow anyone to summon up the pages of books and newspapers, annotate them, and store their work. His great insight was that this would transform the nature of human *memory*—as if the man on the street had been given a photographic mind. Surely such a useful aid to human thought would have to exist someday, wouldn't it? Like a science-fiction author, Bush rocketed past the limits of his own time.

In 1945, he published an essay in the *Atlantic* magazine that would become one of the most impressive feats of futurism the world has ever seen. Bush's essay seated the reader in front of the Memex, which by now had evolved into a desk outfitted with screens that displayed the page of any book or newspaper the reader desired. At that time, the ENIAC—the first general-purpose electronic computer—was still a hush-hush military project; it occupied an enormous cavern and did little more than calculate. So Bush wanted to give the reader the sense of what it would be like to live decades in the future, when the computer revved at the speed of the human mind. "There is a keyboard, and sets of buttons and levers," Bush wrote, describing the control panel on the Memex. "If the user wishes to consult a certain book, he taps its code on the keyboard" and it materializes on the screen. The information people wanted, he realized, should just appear, in the same way that a thought pops into the mind. "Suppose I wish to recall what my Aunt Susie, whom I haven't seen for twenty years, looked like. . . . My brain runs rapidly . . . and suddenly her picture is before my mind's eye. The goal of the Memex is comparable." This idea would echo through the rest of the twentieth century: the Memex, he realized, should let the user jump around in texts, linking ideas and building "trails" of associations—a style that mimicked the associative nature of thought. Bush predicted a personal computer equipped with hypertext, a search engine, display screens, and the Internet. He had done it fifty years ahead of time.

Just as important, Bush had come up with a new idea about how invention could happen: not just with scientific discoveries but also with pure vision. The Memex had been devised entirely in his mind's eye and then captured in words; Bush's vision of the machine was so precise that when *Life* magazine reprinted his article, it included illustrations, as if the machine actually existed.

Bush was carrying forth a centuries-old tradition of diagramming machines that could not yet be built. Leonardo da Vinci, for instance, sketched a human-powered helicopter in his notebook, conceiving it with precise strokes of ink. Of course, a fifteenth-century copter with canvas rotors would not have been able to achieve liftoff, so da Vinci was illustrating a thought experiment—trying out the flight in the laboratory of his own mind.

Vannevar Bush, because he lived in the era of *Life*, could do more than just conceive and diagram futuristic machines. He could share his vision with millions.

"I extrapolated freely as I wrote and implied the existence of various technical elements and devices which were actually then in embryo or even practically impossible," he confided later. "The object was not to propose an actual device, but to try to take a long look ahead." Bush recognized that if he prophesied such a tantalizing machine, future engineers would build it.

Bush's audacious essay did indeed inspire a young engineer named Doug Engelbart to take the Memex further, and to transform the imaginary machine into a real one. Like Bush, Engelbart started with a vision of human potential, and he deployed fantasy and prediction as his tools. It was the grandest and boldest kind of backward engineering. Engelbart's story reveals how much becomes possible when we deploy mental time travel to invent.

The Impossible Is Temporary

One morning in 1950, Doug Engelbart was driving to work at the Ames Research Center, a NASA lab in California. "I'm engaged!" he thought, as the road unspooled before him. He had just proposed to his sweetheart and he felt so giddy that, as he approached the parking lot, he had to make an effort to calm himself down. At that time, Engelbart was responsible for performing mechanical fixes on the wind tunnels in the aeronautics lab. It was a job that many engineers would have coveted, but suddenly, in a jolt of self-awareness, he realized that he was bored by his enviable career. That day, he vowed to quit so he could find a crusade and devote his life to it.

At first he imagined setting off for the tropics, where he would battle malaria among people who were too poor to afford medical care. Then he reconsidered: he had read that when malaria receded, the population would explode, and people who had survived the disease would die of starvation. How could he—or anyone—hope to fight famine or disease when all the world's problems were snarled together?

As he puzzled over this, an idea began to take shape. Engelbart had read Vannevar Bush's description of the Memex years before in *Life* magazine, with its fanciful illustrations of a desk filled with gears and sprockets and two "viewing screens" sitting atop it.

Now the Memex popped into Engelbart's thoughts as the answer to all his questions and doubts. In the same way that he'd just decided to marry the woman he loved, he'd also decided to marry himself to this mission of making the Memex into a reality. If the world's problems were too complicated for the unaided human mind to solve, he would invent a way to become smarter.

You couldn't call it an "aha" moment, because the young engineer hadn't invented anything yet; he'd never even touched a real computer. In the early 1950s, there were still only a few electrical brains in existence, and Engelbart had not had a chance to work with one yet. But his lack of expertise ended up being a great advantage; he hadn't yet glimpsed the tangle of wires and vacuum tubes that it took to execute calculations back then. Instead, his idea of computing pinged off of another visionary's vision; you might call it a fantasy squared, a dream that spun out of a dream. As Engelbart elaborated on Bush's original idea, he imagined a cockpit rather than a desk, one that would allow people to fly around through a virtual space, and a system of thought that you could navigate like a driver zooming through a city.

When he was thirteen, Engelbart had tinkered with a broken-down Ford and had eventually coaxed it back to life. "A car in those days—you used the steering wheel, a choke, and both of your feet on the pedals. Your whole body was involved," his daughter Christina Engelbart told me. She thought that her father's experience with that car inspired his insights about the computer; he realized it had to be drivable by anyone—a secretary, a politician, a scholar.

After he dedicated himself to his crusade in the early 1950s, Engelbart knew he would have to immerse himself in the technology. As a graduate student in Berkeley, he found his way into the most advanced computer lab in Northern California: the CALDIC project, sponsored by the navy, deployed a vast array of vacuum tubes and a magnetic drum to creak through calculations. It bore no resemblance to the Memex at all, and yet it helped him to gain his first big insight about the path forward.

Computers, Engelbart decided, were fundamentally different from the aeronautics equipment he'd worked on at the Ames Research Center. Airplanes, for the most part, stayed the same size. After all, you couldn't seat a hundred passengers in an aircraft the size of a thimble. But ever since the digital computer had been born, it had been shrinking; and with miniaturization came dazzling new abilities, because now electronic signals could zip across tiny landscapes. The cycle of improvement was so frenetic that if you designed a computer and then spent five years building it according to your original plans, it would be obsolete by the time you were finished. So the trick was to recognize that whatever seemed impossible today was only temporary. To be an inventor in this field, you *had* to be a visionary.

Engelbart wandered around the Berkeley campus, searching for fellow travelers who agreed that computers were about to transform society. When his engineering buddies scoffed, he tried out his ideas on the artsy crowd. "The English Lit majors were very eloquent," and fun to chat with at parties, he remembered later. But when he shyly suggested that computers would one day help them understand literature, his new friends would sidle away to refill their wineglasses.

One day he met with the head of a Northern California engineering school to ask for work. In a blitz of enthusiasm, Engelbart described his vision of superpowered computers that would partner with human beings, but before he could finish his spiel, the older man shut him down, saying, "If you continue talking about this, you're not . . . going to last."

By the early 1960s, Engelbart had found a home for his project. At Stanford Research Institute (SRI) he now had the freedom and the funds to build his ideal machine. There was just one remaining impediment: many of the components that he'd imagined did not yet exist. So he made do: he bought one of the fastest small computers then available—it cost more than $100,000—and wired it up to a round monitor the size of a dinner plate. The computer stored instructions on paper rolls. Nonetheless, he managed to use it to hack together a workstation; as he typed words, they appeared on the porthole-like screen, made out of shimmering light. He invented the *experience* of word processing, even before computers were ready to process words. His workstation was achingly slow, and yet it gave the user the feeling of flying around in a new kind of virtual world—what would become known as cyberspace.

At one point, Engelbart even rigged a helmet-mounted pointer so that typists could move the cursor with a nod of the head, but people complained that it gave them a sore neck. In the end, the best cursor-control device turned out to be a box on wheels that the user rolled around the desk like a toy car, which Engelbart had designed himself. The researchers in his lab nicknamed this device the "mouse."

By the mid-1960s, Engelbart had produced a rough approximation of a Memex equipped with a word-processing system, a mouse, a what-you-see-is-what-you-get display, and rudimentary Internet-like functions with hypertext links—in other words, most of the fundamentals of modern computing. But few people were ready to follow him into this future. He was Silicon Valley's own Cassandra, doomed to predict (and even build) the machines of the future, and doomed to be ignored. To many of his contemporaries, his prophecies sounded like gibberish.

In 1968, Engelbart and his engineers from SRI put on a lavish demo choreographed like a Broadway production; he hoped that by showing off the miracles that he had made real, the industry would embrace his way of thinking. Up on the stage, Engelbart narrated his own actions while he edited a document displayed over his head on a screen; he showed how easily he could manipulate the cursor with the mouse; and, most impressive, he communicated with another machine on a rudimentary Internet. About a thousand technology experts crowded the room to witness what became known as the Mother of All Demos.

Though the people in the audience leaped to their feet afterward and gave Engelbart a standing ovation, the demo was met with incomprehension. Bill Paxton, who worked on the team at SRI, said that 90 percent of the engineering community regarded Engelbart as a crackpot. "It's hard to believe now," Paxton told a reporter in 2008, "but at the time, even we [Engelbart's fellow researchers] had trouble understanding what he was doing."

In the 1960s, most people thought of a "computer" as a hulking machine that churned through actuarial tables in the basement of an insurance company. So when Engelbart demonstrated how the computers of the future would help perform secretarial work, his audience was baffled. Why was he turning the computer into a glorified typewriter, when it should be calculating the gross national product of Brazil? They applauded his vision, and then many of them dismissed it.

But a few people in the room that day did grasp the significance of the demo. One of them was Alan Kay, an engineer who sat in the crowd shivering with a high fever. He had dragged himself out of a sickbed because he'd been intrigued by the SRI project, and now as he watched Engelbart, he was electrified. "It was one of the greatest experiences in my life," he recalled later. "Engelbart was like Moses opening the Red Sea." Soon, at Xerox PARC, Kay and his colleagues threw themselves into building faster, smaller, and more friendly versions of Engelbart's visionary prototype.

In the 1970s, engineers at the Xerox PARC lab embraced a new kind of imaginative work, halfway between invention and forecasting. Alan Kay later came up with a name for the new method they pioneered back then: he called it the Wayne Gretzky Game, after the legendary hockey player who skated to where the puck was about to go, rather than to where it had been.

The Wayne Gretzky Game went like this: You would try to imagine the world decades in the future. And then you would figure out what kind of technology was destined to exist, based on what you'd predicted about the evolution of machines and about human needs and desires. You would create sketches, videos, and stories about that technology, in an effort to bring it to life. Alan Kay became a particular proponent of the Wayne Gretzky Game. The secret is "to dream about things that have no connection to the present," he explained in 2014. "The dream can be expanded into a vision, and the vision can combust into ideas that are not like the ideas that are around us." He argued that there was a problem with problem solving: "Anything that is an obvious problem is a manifestation of the current worldview" and prevents you from leaping forward. But if you play the Wayne Gretzky Game, you imagine "something that would be cool to have, or important to have," rather than a solution to a problem. "So one of the things I thought of in 1968 was that it was inevitable that we were [going to] have laptop and tablet computers."

The Wayne Gretzky Game—along with the lab's ample funds and benign neglect from management—allowed Xerox PARC to become one of the most productive invention factories of the twentieth century. One team in the lab built what might be considered the first true personal workstation aimed at the consumer market; called the Alto, it was outfitted with a graphical user interface, word processing, a mouse, and a file-storage system.

In 1977, Xerox executives made one of the famously bad moves in business history. Instead of going forward with the Alto—which might have put Xerox at the front of the personal computer industry—they killed the project. Lots of smart thinkers have developed theories about why the company's managers made this terrible mistake. Often, Xerox's mistake is diagnosed as a decision-making error, as if the managers weighed all the options and carefully considered the Alto, and then ruled against it out of an excess of caution. But "decision making" implies that management was aware of the Alto, that these men had examined the machine, and that they understood it. In fact, the machine may have been invisible to them, hidden in plain sight. One anecdote shows just how difficult it could be at that time for a male executive to even get near a computer with a keyboard.

In 1977, Xerox Corporation invited its executives and their wives to a "Futures Day" conference in Boca Raton, Florida. The event included a room set up with Altos where visitors could play with the machines. Charles Geschke, who was then a researcher at Xerox PARC, remembered that executives—all of them men—treated the machines as if they were radioactive, eyeing them from a safe distance. "What was remarkable was that almost to a couple, the man would stand back and be very skeptical and reserved, [but] the wives—many of whom had been secretaries—got enthralled by moving around the mouse, seeing the graphics on the screen and using the color printer. The men had no background, really, to understand the significance of it," Geschke reported. With the enormous divide between the genders in that era, many objects seemed to be endowed with a gender too—there were "male" things and "female" things. Typing machines belonged to women and computers to men. So asking an executive to sit at a keyboard was like expecting him to cook a pot roast—the Alto might as well have been a frilly apron.

Martin Cooper had no such problem. He'd been able to evangelize for his futuristic phone simply by *handing* it to journalists and congressmen and letting them play with it. (Cooper performed one demo in Washington, DC, and another in New York City.) His audience already understood the phone; just about everyone had felt the frustration of missing calls or spending time tethered to their desks. So as soon as people experienced a phone that would grant them freedom, they wanted it. The cell phone was a magical spin on a familiar item.

But personal computers were a different matter. Few people in the 1970s had ever touched a computer. This made it exceptionally hard to entice most consumers to trust the machine as a personal assistant and a friend. And yet, there were some Lead Users and early adopters who suffered from just the right set of frustrations to recognize the potential of the personal computer. Secretaries gravitated to the machines, welcoming relief from the drudgery of correcting typos. And another group had also become passionate advocates: hobbyists who played with home computers. Once you had programmed a machine like the Altair 8800, with its baffling switches and blinking lights, you would inevitably fantasize about ways to make the job easier. These hobbyists were quick to grasp that in order to jump into the future, the computer would have to go "backward" and become more like a typewriter.

One of those hobbyists, Steve Jobs, toured Xerox PARC lab in 1979. In a now-famous moment, Jobs spotted a desktop computer with a mouse, a monitor, and word-processing features. According to legend, he hopped around and yelled, "What is going on here? You're sitting on a gold mine! Why aren't you doing something with this technology? You could change the world!"

Like the other visionaries profiled in this chapter, Jobs recognized that the computer was not a fixed entity existing in one moment of time. It was evolving—and fast. Cars and vacuum cleaners and helicopters did not change much over the span of a year. But computers were always shrinking. As they did, they fundamentally changed their abilities and their relationship with human beings; the "computer" was a trajectory, an arrow pointed at the future. To understand computing, you had to be able to see the blur of change. You had to be able to deploy a particular kind of imagination.

It was the kind of thinking that Doug Engelbart had used to springboard far ahead of the curve.

Silicon Valley Soothsaying

In the 1950s, Engelbart came up with a profound idea that helped him to predict the future. The concept, which he called "scalability," had to do with the way that small machines could vastly outperform large ones. "What if all of us, and everything in this room, were to become ten times larger?" he liked to ask his audience when he delivered lectures. He would point out that at ten times your current size, you would weigh a thousand times your current weight, so the chair you were sitting on—even if it also grew ten times larger—would collapse beneath you. And once you fell to the floor, you'd lie there like a baby, helpless, because your muscles could not support you. "A mosquito, as big as a human, could not stand, fly, or breathe," Engelbart pointed out. In other words, when you scale up machines to huge sizes, they require far more structural strength and power. On the other hand, if you can build a tiny machine, it acquires miraculous capabilities. The same computer rendered at half the size will become exponentially faster and smarter.

Of course such mind-blowing concepts were the common coin of engineers and physics professors. Engelbart's genius was to add a new twist. He recognized that the weirdness of scalability had huge implications for the future of computing. By predicting the *size* of computer components in five or ten years, you might also be able to forecast how they'd be used and the new possibilities that would open up. But he had trouble making other people see the implications of this principle until one day in the early 1960s, when he delivered a lecture and Gordon Moore rushed up afterward, eager to continue the conversation. Moore was an engineer at Fairchild Semiconductor, a company that made a new kind of integrated chip packed with tiny circuits, and he was about to turn Engelbart's idea into one of the most accurate forecasting methods that the world has ever seen.

• • •

In 1965, Gordon Moore received a strange invitation: "I think you might have fun doing this," read the letter from Lewis Young, the editor of *Electronics* magazine; he suggested that Moore forecast what the information-processing industry would look like in a decade. "I find the opportunity to predict the future in this area irresistible," Moore wrote back, and soon his manuscript followed. That article, which Moore tossed off within a few weeks, turned out to be one of the most important documents of the late twentieth century. Thinking back to Engelbart's lecture on scalability, he realized that the concept could be turned into a method for seeing the future. If you could predict the *size* of integrated circuits a year from now, you could predict their speed and power.

Moore pointed out that information-processing circuits had doubled their performance every year for three years in a row, and he projected that trend would continue steadily for the next decade, until 1975; if airplanes were to improve at the rate that Moore had predicted for silicon chips, they'd soon be flying ten thousand miles an hour. With this notion, Moore remembered later, "I was just trying to get across the idea that integrated circuits were going to be the route to cheap electronics, something that was not clear at the time."

By 1975 the computing power was almost exactly as he had anticipated. "I had no idea this was going to be an accurate prediction," he said later. His observation became known as Moore's Law. Though he certainly did not mean to suggest that it would hold for all time, like a law of physics, it became something even more powerful than a natural law: it became a way to make money.

A corporate planner could consult Moore's graph to estimate the cost and power of computer chips and then decide what kind of machine to build within a year or five years. Soon thousands of managers at companies all around the world were using the same playbook. Though Moore's Law wasn't intended to be a rule of centralized planning, it has functioned that way. It created a sense of predictability, and that in turn funneled trillions of dollars in investment into the computer industry. No one is in charge of Moore's Law; no one legislated it into existence; and yet it "is Silicon Valley's guiding principle, like all ten commandments wrapped into one," as the tech visionary Jaron Lanier has written.

In truth, there was no one single Moore's Law. As the chip market slowed down a bit, so too did Moore's Law—today's version predicts a doubling of power every eighteen months, rather than every year. Nonetheless, the law (or laws) has reached into every corner of our designed environment and channeled a fortune toward Silicon Valley.

What Can We Predict?

In the mid-1960s, a military strategist named Herman Kahn launched a grand experiment in futurism. Working with his coauthor, Kahn asked a team of powerful men—corporate managers, military brass, policy wonks—to describe what the world would look like in the twenty-first century. The result was a book titled *The Year 2000.* It included a detailed list of one hundred predictions about the technologies that would exist thirty-three years in the future.

After the actual year 2000 passed, a researcher named Richard Albright became curious about how many of the technological predictions had come true. He thought the success rate might reveal something about the reliability of forecasts in general. And, ingeniously, he decided to break the book's one hundred predictions into categories—medical, leisure, transportation, and so on. He was interested in whether certain *kinds* of breakthroughs are more predictable than others; the list of one hundred technologies gave him an unusual chance to answer that question.

In fact, Albright discovered an enormous variation in the predictability of different fields. The forecasts related to computing and communication achieved a dazzling 80 percent accuracy rate. That is, Kahn's experts had been able to anticipate and describe the Internet, VCRs, and cell phones with a precision that seems to defy logic.

But the experts were terrible at forecasting other kinds of technology. The worst guesses clustered around the fields of medicine, architecture, and transportation. Back in the 1960s, the think-tank experts had imagined us living in undersea cities and enjoying "programmed dreams" while we slept. They predicted that we would hibernate aboard spaceships bound for other planets. We would take pills to control our appetite—meaning that no one would get fat. Doctors would possess a new arsenal of drugs that would cure genetic diseases. And, of course, cars would fly.

Why was it possible to predict the future of information technology but little else? This is a hotly debated question, and the combatants fall roughly into two camps. Those of a teleological bent see a grand design at work. For instance, Kevin Kelly, one of the founders of *Wired* magazine, takes the position that there is something inherent in the nature of information technology that causes it to evolve according to certain laws. "If, in a counterfactual history, communism had won the cold war, and microelectronics had been invented in a global Soviet style society, my guess is that even that alternative policy could not stifle Moore's Law," he wrote. So as he sees it, Moore's Law springs out of the technology itself. Kelly and his camp argue that economic and social forces are nearly irrelevant—instead, it's the scalability of logic chips (and a few other technologies) that makes them special.

On the other side are the people—including Moore himself—who argue the opposite position. They regard the "law" as something like a marvelous mass delusion. In 2005, Moore remarked that when he proposed his theory in the 1960s, he inadvertently launched lemming-like behavior around Silicon Valley. "All the participants in the business recognize that if they don't move that fast they fall behind technology, so essentially from being just a measure of what has happened, [Moore's Law has] become a driver of what is going to happen," he commented. According to this theory, Moore's Law is like the emperor's new clothes—if all the courtiers agree to believe in the ermine cloak and silk pantaloons, the emperor can parade through the streets clothed only in consensus. Likewise, if lots of companies agree that the computer industry is destined to run on eighteen-month cycles of improvement, this self-fulfilling prophecy will come true.

The bottom line: it's impossible to know why Moore's Law happens. The history of computers can't be sliced out of life in general and examined as a separate entity, like a liver extracted from a lab rat. We can't know what would have occurred in Kelly's counterfactual world where the Kremlin controlled the chip industry. So the only sensible thing to do is move on to a more practical question: What can you *do* with Moore's Law? How can it be used as a tool of invention? Alan Kay, for instance, described the way engineers at Xerox PARC figured out what would be possible in ten or twenty years; to do that, they had to observe patterns of evolution and project them into the future, and certainly Moore's Law became key to that way of thinking.

And thanks in part to Moore's Law, the artists and engineers at Pixar dared to plan for computer animation decades before it existed. Alvy Ray Smith, the company's cofounder, wrote that in the 1970s computers could do little more than sketch lines, but he and his collaborators were confident that the impediments would melt away year by year. "We . . . conceived the notion of the first completely digital movie almost four decades ago. It took 20 years to realize that dream with *Toy Story*, but Moore's Law is what gave us the confidence to hang on for those two decades," he wrote in 2013.

But, of course, a forecasting tool is just one piece of the equipment required to succeed at the Wayne Gretzky Game. You also must deploy that mysterious Vannevar-Bush-ian voodoo, that ability to work in a lab inside your head, to construct a machine that is impossible to build in the present moment; you must be able to dream up every detail, right down to the translucent screens.

That feat requires a muscular imagination, one that is powered by a dynamo of mental focus. You must be able to transform that blurry inner world into a cinema, and then learn how to run the movie backward and forward so that you can test out your concept, or time-travel into the future. The quest to control the imagination is age-old, but it took some particularly strange turns in the twentieth century. In the next chapter, we will follow one renegade engineer who believed he had found the way to harness the inner world and turn it into an invention factory.

10

THE MIND'S R&D LAB

In the early 1960s, just a few miles away from Doug Engelbart's Stanford lab, another engineer was performing experiments that would transform the culture of Silicon Valley, too—but these experiments involved minds rather than machines. Myron Stolaroff believed that drugs could be used to rev up the imagination. In his quest to understand how to enhance the performance of inventors, Stolaroff performed some of the most ambitious studies of the inventive process that have ever been attempted, and his results give intriguing hints about how we construct entirely new technologies in the mind's eye.

The story begins in 1956, when Stolaroff was an engineer and corporate planner at Ampex, a company renowned for running one of the most productive R&D labs on the West Coast. At that moment, Ampex was hiding a secret in its office: the world's first video-recording machine, which it was readying to debut in the spring. Stolaroff—who'd contributed key ideas to the project—should have been savoring his success. But instead he sweated with dread. "I was slight of build, deadly serious, extremely introverted, trembling at whether others approved of me or not, anxious to follow all the rules and conventions of society," he wrote later.

And that's why Stolaroff had begun working on his own secret project, a drastic plan to reengineer himself. His aim was to form "a satisfactory relationship with God." He had learned about a strange new drug called LSD that was rumored to produce religious experiences. The drug, still largely unknown, was then legal but difficult to obtain.

That spring, Stolaroff hurried off to Vancouver, British Columbia, to meet with Al Hubbard, a Canadian who ran hush-hush experiments with drugs in apartments and hotel rooms. During that LSD trip, Stolaroff experienced a religious awakening and deep sense of communion with the divine—he did find God. Strangely enough, he also discovered what he thought would be a key business opportunity, one that would give Ampex an edge over its competitors.

Stolaroff was a talented engineer and something of a visionary when it came to finding new possibilities. In the 1940s, he had designed the electronics inside the tape heads in the Model 200A magnetic recorder—the legendary machine that Bing Crosby used to capture his live shows. Stolaroff had also schmoozed with rocket scientists at White Sands and realized that they could use magnetic tape to record the data from their launches—opening up a new market for Ampex.

Stolaroff was always one to recognize an opportunity that would give his company an edge in the market. And so now, after his first taste of LSD, he became convinced that the drug could focus the mind's eye. With it, he believed, the engineers at Ampex would be able to solve technical problems that had stumped everyone else in the industry. "In the LSD state, it is possible to reach levels where the mind is sharp and clear," he wrote later. "Fresh ideas and perspectives flow unhindered, presenting many new possibilities, often of great value." The drug was "the greatest discovery that man has ever made," a master key to the imagination, he decided. Stolaroff believed he could use LSD as a rocket fuel for the inventor's mind.

If this sounds like an odd way to think about drugs, it was also an unusual way of looking at engineers. In the 1950s, they were still the guys in the workshop with their shirtsleeves rolled up—the descendants of Edison's "muckers" who carried out orders and kept screwdrivers in their back pockets. But Stolaroff had recognized that a new era was coming and that a new kind of imagination would be required. The scientists and engineers who could create the most vivid mental pictures would be the ones who designed the future.

Of course, Stolaroff's quest—to control the mind's eye—was hardly new. It began as far back as 1400 BC, when a temple in Delphi became famous for its fortune-telling priestesses. According to legend, these women descended into an underground vault to inhale the fumes that rose up from cracks in the floor. In 2002, a group of researchers published a provocative paper in which they showed that the Delphi temple was indeed positioned over two fault lines; centuries ago, gases from petrochemical deposits below the floor of the temple would have swirled up into the chamber. The fortune-telling drug appears to have been ethylene, which induced hallucinations and inspired the priestesses' predictions.

But drugs were only one way to enhance the imagination. Mathematicians, scientists, and astronomers trained their powers of concentration, practicing until they could conjure up worlds in their minds. In the sixteenth and seventeenth centuries, when Galileo set out to prove that the Earth rotated around its axis, the Italian physicist concocted mental scenes worthy of a fantasy novelist. In the experimental laboratory of his own mind, he built an oceangoing ship and then climbed down a ladder into the rooms below deck; there, butterflies danced about in the rafters, winking the dim light that filtered in from the portholes. Galileo had put those imaginary butterflies into an imaginary ship to prove a point; just as those creatures flitted around without being aware that they were on a moving vessel, he argued, so too did humans walk the Earth unaware that the planet was whirling around in space. At the time, imaginary scenarios could be used to argue for scientific concepts.

By the nineteenth century, thought experiments were so familiar that the method was sometimes taught to schoolchildren; as a teenager at a progressive school, Albert Einstein learned the principles of mental experimentation, and at the age of sixteen, he started to wonder what it would be like to zoom alongside a beam of light. As an adult, of course, he famously used his imagination to work out his ideas in a landscape he outfitted with hurtling trains, locked trunks, stopwatches, elevators, and beetles.

Nikola Tesla, the visionary inventor, also used a lavishly outfitted mental laboratory. As a boy confined to his sickbed, lying in the dark waiting for sleep that would not come, Tesla taught himself to visualize elaborate worlds that did not exist: "Every night (and sometimes during the day), when alone, I would start on my journeys—see new places, cities and countries—live there, meet people and make friendships and acquaintances," he wrote later. "This I did constantly until I was about seventeen when my thoughts turned seriously to invention. Then I observed to my delight that I could visualize with the greatest facility. I needed no models, drawings or experiments. I could picture them all as real in my mind. Thus I have been led unconsciously to evolve what I consider a new method of materializing inventive concepts and ideas." Tesla touted this mental workshop as a place where he could instantly test out and perfect his machines: "Before I put a sketch on paper, the whole idea is worked out mentally. In my mind I change the construction, make improvements, and even operate the device."

It was only natural that inventors would dream of a technology that could turbocharge the mind's eye. Tesla hoped to invent a camera that could take pictures of the world inside our imagination. "'I expect to photograph thoughts,' announced Mr. Tesla calmly," according to a newspaper reporter in 1933. An illustration that accompanied this newspaper article showed an inventor communicating telepathically with a machine that looked like a slide projector, so that his mind's picture glimmered before him on a screen.

And the dream continues. Elon Musk—the inventor who upended the car industry—has built a system that lets you bend and mold the design of an engine part by waving your hands like a conductor, and then print out the object. "I believe we're on the verge of a major breakthrough in design and manufacturing, in being able to take a concept of something from your mind and translate [it] into a 3-D object intuitively on the computer," he said in a video demonstration.

Years ago, when I was a graduate student, I had a dream about my own imagination-boosting machine. Back then, I was struggling to finish a novel, and I frequently felt that the sentences turned gray and fusty when they emerged from my fingers, and the story kept fizzling out. One night I collapsed into sleep and dreamed about a computer circuit board made of green plastic covered in gold threads and bejeweled with capacitors that looked like pieces of candy. I realized that I only needed to open the hood of my computer, install this magical circuit board, and it would do all the work of turning my vague story line into vivid Technicolor fantasies that I would capture in words, and then a work of genius would fly out of my fingers. When I woke up and realized the circuit board didn't exist, I felt bereft—there, on the other side of the room, my computer squatted, fat and gray and stupid. And I would have to spend the day banging into it again, getting lost in my inner world, frustrated and failing.

But of course, in some way I *did* possess the magic circuit board, as we all do. Our brains are designed to simulate reality, allowing us to test out wild ideas. At the same time, all of us have struggled with the feeling that our mental equipment just isn't up to the task we hope to accomplish.

And that's why Stolaroff was so excited. He believed he had discovered what so many people through the centuries have sought: a booster for the mind's eye.

After his first acid trip, Stolaroff rushed back to California to share his revelation with his fellow managers at Ampex. At a corporate meeting, he raved to the board of directors and the president that he had found a secret weapon that would put the company at the top of the electronics field. The plan was this: Ampex would hand out LSD to its engineers.

One can only imagine the reaction of the men in that room. Someone in the meeting pointed out that LSD was a relatively untested substance that could damage the human brain. And why would Ampex need to boost the genius of its engineers anyway? The company was besieged with orders for the Mark IV video recorder, and money was rolling in. The managers suggested that Stolaroff calm down, forget about his LSD trip, and go back to work.

But when he described the effects of LSD to the engineers themselves, they jumped at the chance to try it. At a cabin in the Sierra Nevada, eight Ampex engineers gathered to take part in Stolaroff's experiment. They tripped on LSD and focused their minds on blueprints and circuits. According to Stolaroff, several of them felt that they made creative breakthroughs during the LSD session. Armed with this "evidence," Stolaroff again presented his case to his fellow managers at Ampex. They were not impressed.

This was a classic outsider-versus-insider moment. Stolaroff was vibrating like one of his own antennas, picking up emanations from a psychedelic revolution that was just then in the offing. His plan for the company back in the 1950s—to dose the engineers on drugs—was one of the more crackpot schemes in the annals of capitalism. But he'd been prescient, anticipating the rise of a new type of engineering that involved artistry and a science-fiction writer's imagination. In the 1960s, as Moore's Law took hold in Silicon Valley, you had to do more than engineer a great product; you had to live in the future, to be able to guess what would be possible in ten or twenty years. Imagination—and particularly that feral, freewheeling variety that didn't thrive in captivity—would become a sought-after commodity. To succeed, you would have to be as oracular as the Delphi priestesses. You would have to inhale the fumes.

Still, Stolaroff's bosses at Ampex wanted no part of it.

And so, wrote Stolaroff later, "in 1961 I resigned from Ampex, set up a non-profit corporation which I boldly and naively named the International Foundation for Advanced Study, and located offices and research space in the town of Menlo Park, California." He outfitted his lab with Persian rugs, eyeshades, headphones, drawing pads, and an exquisite stereo system.

"Each morning I would come to work and park in the Menlo Park downtown parking lot which adjoined our building," Stolaroff wrote of those early years of the 1960s. As he strolled over to the door of his office, he snickered at the shoppers who parked their cars nearby; how could they fail to notice what was going on in the suite above the beauty parlor? There, some of the top creatives in Northern California dropped LSD and mescaline, and it was all perfectly legal in that era before government regulation of psychedelics. "There was little curiosity in our work," Stolaroff wrote, miffed that his revolutionary experiments did not cause more outrage.

He was able to recruit engineers, physicists, and computer pioneers floating around the Stanford campus. Stewart Brand was just one of the luminaries who tasted psychedelic drugs under the guidance of Stolaroff. So too was Douglas Engelbart. And also Irwin Wunderman, an engineer at Hewlett-Packard nicknamed "Mr. Transistor." Many of them had been drawn in by their desire to do better work or break through technical problems that had stumped them.

Stolaroff had assembled an unusual team of experts to run the psychological experiments. For instance, one of Stolaroff's chief collaborators, Willis Harman, was an electrical engineer who worked as a long-range planner at Stanford Research Institute. At SRI, he had supervised a group of "futurists looking at the changes that were taking place in the world, to help corporations and government agencies with their planning," Harman told an interviewer years later. "After we'd been doing this for a couple of years, one of my staff came to me and said he had to resign because he couldn't stand . . . to come in day after day and look at the future." Like Stolaroff, Harman had realized that prediction and "mental R&D" required exhausting focus, and he too wondered whether these abilities could be boosted in the human mind.

The experiments they designed were "to see if you could use the lens of a psychedelic to tightly focus down on a scientific problem," according to Jim Fadiman, who helped conduct the experiments when he was a graduate student in psychology at Stanford. "We brought in senior scientists from a number of the local industries that were developing a lot of different products."

The volunteers were given cocktails of drugs throughout the daylong session—not just LSD and mescaline but also amphetamines and an antianxiety drug called Librium. In retrospect, it seems amazing that the volunteers could function at all on this brew, but they did.

During one of the morning sessions, the facilitators challenged a group of volunteers to invent a phonograph system that would also address the problem of wear and tear on the vinyl records, which deteriorated a little bit every time they were played. As the men pondered this problem, they listened to music on a stereo system—the very sort of machine they had been asked to redesign. At one point, the needle hit an imperfection on the vinyl LP, causing a crackle of distortion. One of the men became particularly upset by the noise. The man—code-named Member H by the researchers—told the group that he had been imagining himself as the needle, riding in between the grooves of the vinyl record, trembling with the vibrations that made the music—and it had been *awful.* He had been so upset by the needle grinding into the record that he had decided to turn himself into a tiny airplane instead. Now, he announced, he was flying above the surface of the vinyl record as it spun around. And it was great! He felt so much better!

And then another of the volunteers in the room, code-named Member G, suggested an idea: What if the record-player needle could be made of a material that would not scratch the vinyl? Picking up on Member H's fantasy, he suggested using a beam of light to read the music encoded on the surface of the record. This was a wildly inventive idea. Laser technology had been discovered just a few years before. While the beams of light were used in laboratories, they had not yet made their way into consumer products. But now a band of psychedelic seekers had hit on an idea that was poised to transform the recording industry—the laser could be used like a record-player needle. The men discussed how a beam could be designed to detect signals and some of them proposed schemes for transmitting the information in pulses of light.

The report on this session was issued in 1965. A year later a physicist named James Russell filed a patent for the first optical-digital recording and playback system. Russell's revelation had been remarkably similar to the thoughts voiced by Member G.

The men who participated in the study were most enthusiastic about the afternoon sessions, when they were allowed to work on their own. For three hours, during what the researchers called the "illumination phase," each man would meditate on a problem that had stumped him.

Twenty-seven volunteers participated in the illumination-phase experiment, in which they were tasked with producing a novel invention or design; ten of the men reported that they had solved a problem, and most of the others had made significant progress toward an answer. Several inventions resulted. One of the volunteers in the study figured out an improvement for the magnetic tape recorder; another redesigned a microtome (a machine that slices samples so that they can be examined under the microscope); and yet another figured out how to steer the electrons inside a particle accelerator, according to the authors of the study. Other creations included the architectural plans for an art center and a new theory about photons. Just as Stolaroff had anticipated, the psychedelic sessions produced impressive results. But unfortunately, he and his team of researchers neglected to run tests on a control group, so we have no way of knowing what would have happened if some volunteers had been given a placebo rather than a cocktail of drugs. It might be that this talented group of engineers and designers would have performed just as well even when sober.

The question about LSD and creativity has remained unanswered for five decades. After Stolaroff's experiments, it became nearly impossible for scientists to re-create his research. A 1970 ban in the United States made it illegal to test the drug on human subjects, and international laws prevented experiments in other countries.

But that has begun to change. In the past few years, scientists at a few institutions have received the go-ahead to perform clinical trials with psychedelics. In 2015, as I was finishing up this book, a team based at Cardiff University in Wales had undertaken a sort of reboot of the Stolaroff experiments; the researchers are studying the effects of LSD on volunteers to find out (among other things) whether the drug facilitates creative breakthroughs. The volunteers were being dosed up, put into brain scanners, and then challenged to perform a barrage of thought experiments. As I write this, the results have yet to be published. One of the researchers involved with the study, Robin Carhart-Harris, asserts that "understanding the brain mechanisms that underlie enhanced cognitive fluency under psychedelics may offer insights into how these drugs may be psychologically useful, for example, in helping patients experience an emotional release in psychotherapy or potentially enhancing creative thinking."

Of course, whatever the findings of the study, artists and poets and inventors will continue to search for drugs to heighten the imagination. For centuries, people have dosed themselves with chemicals in an effort to run faster and leap higher, and they have also taken performance-enhancing drugs to tweak the imagination. We have always wished that our mind's eye could be more powerful. The number of substances people have ingested in an effort to focus that inner eye—from belladonna to tree bark—attests to the keenness of that desire. It can be exhausting and defeating to struggle with our own imagination, and it's no wonder that so many have sought a shortcut.

It's common to assume that the imagination is a joyful theme park, the Six Flags of the mind. We talk about losing ourselves in daydreams or retreating into fantasies. But many people find it painful to spend more than a few moments in that inner world. A recent study revealed that about a quarter of the female subjects and two-thirds of the men preferred to endure electrical shocks rather than sit alone with their own thoughts for fifteen minutes. For many people, the imagination is like the summit of Mount Everest, where the view is transcendent but the air is too thin to breathe.

As a longtime writing teacher, I've been privileged to watch as college students learn how to explore this inner world, suffering its discomforts in order to mine its treasures. One student I worked with years ago, a football player who could quote lines from *The Godfather*, arrived at our first class bragging that he had dreamed up the plot for an entire movie. "All I have to do is get it on paper," he told me.

A week later, he dragged himself into my office and slumped into a chair. "I have the whole thing in my head. But as soon as I try to write—*grrrrr*." He clenched his fist to show how the words snarled up.

I knew what was happening. He *felt* the story; he heard the music of it playing inside him; but he didn't know how to proceed with the painstaking work of actually building up an entire world and bringing characters to life. "Tell me: how does the story open?" I asked.

"A car crash," he said.

So I took him through all the questions I would have asked myself. Were we looking down at the crash from a police helicopter? Or were we seeing the scene from the point of view of one of the victims lying on the ground? What odors hung in the air? Had the ambulance arrived, and if so, was the scene lit by the flare of red and white lights? To tell a story cinematically, you must figure out hundreds of details like that.

And yet, it can take a heroic effort to conjure up even the most rudimentary mental simulation. Simply rotating a cube in your mind can be an exhausting feat. You can fact-check what I'm saying for yourself right now. Imagine a cube balanced on one of its corners and try to turn it a few degrees, picturing how it looks as it moves. When I do this, my cube "jumps" into the positions that are easiest to envision, but mostly it winks out of existence. And as I attempt this feat, I feel cranky and inadequate. You will also notice that it takes considerable effort, a kind of *willpower*, to construct a picture in your mind.

It's just as effortful to think about the future. Look around the room where you're sitting right now and try to "fast-forward" it by fifteen years. To do this properly, you have to ask yourself innumerable questions: Is the air outside still breathable? How do people communicate now? Does America still exist? And on and on. You will notice that as you begin working out the details of your future, you are telling a story. The R&D lab inside our minds has a highly *narrative* quality to it. If you're inventing a futuristic machine, you have to tell a story about the people who will use it: Where do they live? What are they worried about? What do they desire? In the next chapter, we will explore why works of science-fiction literature and cinema have yielded technological ideas, and how futuristic storytelling can be harnessed as a tool of invention.

11

HOW TO TIME-TRAVEL

FOR MORE THAN A CENTURY, SCIENCE FICTION AND INVENTION have been caught in a kind of quantum entanglement; certain technologies have emerged from a lab and simultaneously from the pages of pulp novels. "Extravagant Fiction Today—Cold Fact Tomorrow"—that was the motto of the early-twentieth-century magazine *Amazing Stories* that helped readers to peer over the horizon at wonders like the television. Hugo Gernsback—the magnate who published a stable of sci-fi magazines—argued for a new storytelling genre that he called "scientifiction," halfway between engineering and fiction. Side by side with his magazines about time-travel machines and starships, he released journals for inventors and put Nikola Tesla on the cover of one of his publications.

Indeed, hundreds—perhaps even thousands—of technologies have begun as stories or movie props and then tiptoed into reality. In the original *Star Trek* TV series, the medical tricorder looked suspiciously like it was made out of a cassette player and a salt shaker, but it functioned in a mind-bending way that infected our fantasies: it could peer through the skin and deep into the body, scanning and sensing to come up with an accurate diagnosis. *Star Trek*'s creator,

Gene Roddenberry, in an effort to inspire real-world engineers, reportedly inserted a clause in his contract with Desilu/Paramount that would let anyone who created a hand-held, computerized diagnostic tool call it a "tricorder." He intended his *Star Trek* gizmo to act as a kind of prototype, and indeed it did. A number of companies are now racing to create real-life versions of the machine.

We're used to the idea that a science-fiction story can predict or help create a demand for a new technology. But less discussed is this: inventors often do the same kind of imaginary work. Brian David Johnson, the futurist at Intel Corporation, advocates for experimenting with written narratives, films, and cartoons to envision new possibilities—a technique he calls "science-fiction prototyping." We can "use these fictions to get our minds around what that thing might one day be," he wrote.

As we saw in the previous chapters, Silicon Valley embraced the idea that the engineers in polyester shirts and pocket protectors could double as dreamy soothsayers. As computer chips, and Moore's Law, spread into many other industries, so too did science-fiction thinking.

Perhaps no one has ever deployed that kind of thinking as artfully as Genrich Altshuller, a popular Soviet science-fiction writer of the mid-twentieth century who set out to reinvent invention itself. Altshuller spent most of his life in the mountainous region where Europe melts into Asia. Mentally, too, he inhabited a borderland. He was one of the first people—perhaps *the* first—to create a formula for invention that involved science-fiction-style prediction, cognitive science, and deep knowledge of the way technological systems (like aviation) progress through history. Though bits and pieces of Altshuller's method eventually caught on in the corporate world, few Americans are aware of his existence. And that's an oversight I'd like to correct. In the 1950s, Altshuller declared a new *science of inventing;* and he recognized, before perhaps anyone else, that it was possible to study how machines arise in the human mind. He should be considered the father of Inventology. And yet, many of his writings still have not been translated into English. So in this chapter, we'll try to understand the groundwork that he laid out for the new discipline he had invented.

As a boy in the 1930s, Genrich Altshuller wandered a cobblestoned maze of streets in Baku, a city with views of the Caspian Sea and oil derricks strewn around like children's toys. These, along with the books of Jules Verne, fed his imagination. By the age of fourteen, he began to draw plans for his own inventions inspired by the fictional technologies in *20,000 Leagues Under the Sea*. "Captain Nemo walked on the ocean bed—that's why we need pressure suits," he thought, and then he designed a diving suit.

By the age of twenty, he'd joined the Soviet navy, a skinny kid with a black sailor's cap wedged on his head like a plug to contain all of the ideas that bubbled inside him. The Soviet ships limped through the water, their engines wheezing, always in need of ingenious repair—and Altshuller became the fleet's fix-it man. His wizardry attracted notice, and soon he was working as an inspector in a patent office connected with the Baku naval base, where he pored over stacks of applications for new inventions in aviation, medicine, weaponry, and chemistry. Many of the would-be patents that came into the office were terrible; he'd scan a blueprint and realize that the machine wouldn't work. He began to wonder what separated the failures from the successes. How did some inventors solve problems that stumped everyone else? He and his best friend, Rafael Shapiro, wandered through the library, looking for answers to this question.

They assumed they would find lots of books that taught the craft of invention. After all, "there were whole shelves of books on patent law and patent research, . . . [so] there ought to be ten times as much material on the psychology of inventors' creativity, on techniques for solving inventors' problems," Altshuller said later. Instead, he discovered that almost nothing useful had been written on the subject.

One day, in 1945 or so, it dawned on him that all the secrets to inventiveness had been right in front of him all along, hidden in the patent system itself. He realized that the most valuable clues to the human brain are revealed in the things we design, in the way one machine improves on its predecessor, and that technology evolves over time—in other words, *the patent system is a mirror of the human mind operating at its peak.* It could be mined, like a mountain, for veins of ore; if you only knew how to search through it, you could discover the core principles of creativity itself.

Today, in the age of big data, we recognize the value of this kind of analysis. But in the 1940s, few people had ever tried to data-mine their way to an answer. Altshuller, an Azerbaijani with no phone in his house and limited access to news of the world, had realized the power of large amounts of data to reveal patterns. For two years, he toiled at the patent office and camped out in the library, returning to his parents' home only to drop exhausted into bed, as he struggled through the entire patent system. Later he claimed that he identified two hundred thousand patents that he thought would contain clues about successful problem solving and that he read through forty thousand of them to find patterns.

In the blueprints of the past, he could trace the origins of great breakthroughs and observe how inventors leaped forward—for instance, he might study a patent that showed how one mechanic built an engine into a bicycle in order to create the first motorcycle, and then map out how that idea spread as other minds improved on it. By the late 1940s, he had begun to boil down all that they had learned from the patent system into a set of principles. "Inventors always have two secrets," he wrote in 1961. The first, of course, is the creative insight that leads to a new technology or improvement. "The second is *how* they make their inventions. . . . Sailors have long been mapping the currents, sandbars, and reefs . . . so that everyone knows about [the dangers]. For centuries, inventors have done without a map; every beginner has been making the very same mistakes," he observed. And now Altshuller would chart that map. He believed it would protect the next generation from crashing on the shoals of befuddlement and guide them to open waters where they could discover vast new continents of technical knowledge.

Then, just as Altshuller was actually putting his system into practice to invent novel technologies, he was seized by the military police, thrown into prison, and accused of sedition. In Part V of this book, we will return to this story and examine why he was seen as a danger to Stalin's regime and what that tells us about the politics of inventing. But for now, let's fast-forward to the mid-1950s, after his release from Vorkuta labor camp, when he returned to his home city of Baku and took up again his crusade to understand the nature of invention.

After his release, Altshuller holed up in his apartment and began typing furiously. He was determined to make a living as a science-fiction writer, and by the late 1950s he was succeeding at that, publishing stories under several pen names; he would become one of the best-known Russian science-fiction writers of that era.

He considered his stories to be not just entertainments but also blueprints of machines that could exist in the future. "Since my childhood, . . . science fiction determined my life. It is a kind of religion [for me]," he confided to friends in a 1964 letter. "I prefer prognostic science fiction" he added, because it allows the writer to "look into the future as precisely and far as possible."

In a particularly impressive example of that prognostication—a story published in 1966 and titled "The Donkey Axiom"—he described a working 3-D printer that could create objects out of powder. In the story, he explained that if you look at historical trends, you realize that personal factories (i.e., 3-D printers) will *have to be invented* in the twenty-first century. The cycle of improvement is continually speeding up, he pointed out, and he imagined a time in which conventional, large-scale factories wouldn't be able to keep up with the pace of technological breakthroughs. After "color TV was developed, . . . hundreds of millions of [black-and-white TV] sets in perfect working order were thrown away." Pretty soon, the color sets would become outdated, too, Altshuller predicted, and they would be tossed out to make way for "stereovision." By the twenty-first century, he forecast, products would be obsolete in a matter of days. Then consumers would need to update technology constantly, "mercilessly throwing out billions of new machines simply because they are no longer the latest thing." At that point, products would be made from powder so that they could be continually destroyed and resurrected. It's an unnervingly prescient take on the logic that has led us to where we are today—like Vannevar Bush, Altshuller had been able to construct that new technology in his mind decades ahead of time.

In the 1960s, Altshuller began advocating for a "registry of science-fiction ideas"—it would be something like a patent system for imaginary technologies. He began compiling a list of machines that had been described in futuristic stories and studying how these ideas became real-world inventions. "In the novel *20,000 Leagues Under the Sea*, Jules Verne for the first time expressed . . . the idea of a double-shelled hull for a submarine. The patent on double hulls was issued thirty years later to the French engineer Leboeux." Altshuller pointed out that Verne's novel contained exactly the kind of details that were included in the later patent. So why was it that Leboeux, rather than Verne, received credit for the idea? A science-fiction story, he argued, was really a kind of speculative patent on a machine that *might* exist. "Very often, the ideas of fiction writers are directly used during the early development stage of a new field of science and technology," he wrote. Science fiction "helps overcome psychological barriers on the road to 'crazy' ideas."

He became passionate about taking other people down that road with him. In 1961, he published a book—*How to Learn to Invent*—in which he described his theories and thinking tools. His method would later become known in English-speaking countries as TRIZ (an acronym to designate the Russian phrase for "theory of inventive problem solving"). In the 1970s, he founded an unusual school in Baku where he taught his inventive thinking; and when the Soviet authorities shut down his school, he traveled around the country leading workshops.

Though widely published in the Soviet Union and legendary in sci-fi circles, Altshuller still had no phone in 1975, so the only way to reach him was to show up at his apartment. One day a teenager named Victor Fey did just that. "He was about six feet tall [and] broad shouldered, [with] blue eyes and Nordic features," Fey told me of his first impression of the man who would become his mentor. Fey, then a college student, had come to learn more about Altshuller's system. So Altshuller began to tutor him. "It wasn't formal education," he told me. Altshuller challenged him with technical problems—for instance, what would happen if, instead of building an oil-mining system to operate on Planet Earth, you were trying to perform the same task on a planet with zero gravity? "We would also discuss . . . history, philosophy, psychology, you name it," Fey said.

Fey (now a mechanical engineer and problem-solving consultant based in Detroit) also sat in on Altshuller's workshops that would sprawl over days or weeks. The teacher required his students to read science-fiction novels and generate their own stories—this was a way of encouraging them to use narrative as a "laboratory" for invention. He might ask his students to imagine that they lived on a planet that shrank and expanded every few hours. What kind of climate would it have? What kind of animals would thrive on the planet? Or imagine that diamonds were so plentiful that they could be found in any vacant lot. How would technology evolve? "We all tried to emulate Altshuller as much as we could," Fey said. "We tried to do, to see engineering, politics, culture . . . as part of a much bigger whole."

Disciples came from all over the Soviet Union and allied countries (like Vietnam) to study with Altshuller; they then started their own schools and consultancies, spreading the system throughout the world. In the late 1980s and 1990s, some of Altshuller's followers began developing software packages they had designed in order to guide engineers through TRIZ exercises and imaginative work. As it spread, TRIZ would eclipse its originator, and many people who encountered the system knew little about Altshuller or how he had originally taught his ideas.

Victor Fey pointed out that his teacher "was a science-fiction writer with a very keen interest in developing human imagination." Altshuller's own imagination was so capacious that he could construct a whole system in his mind and then "fast-forward" it so that it evolved over time, according to Fey. That way of thinking could not always be communicated in a book or software package. TRIZ was only a shadow of the mind that made it, and as followers spread the system, it became a copy of a copy of a copy, especially when translated into awkward English.

Nonetheless, in the late 1990s and early 2000s, TRIZ became something of a corporate darling and migrated into dozens of Fortune 500 companies from Amoco to Motorola to Xerox. The Russian creativity method "is fast becoming the innovation 'it' concept around the globe," wrote business reporter Andy Raskin in 2003.

Like a true tinkerer, Altshuller never stopped fiddling with the TRIZ formula for creative problem solving, and it evolved through the decades into a vast Rube Goldberg machine of thought; his followers refer to TRIZ-56, TRIZ-59, TRIZ-64, and so on, to indicate all of the different versions of the system as it changed year by year. It involves a labyrinth of concepts, big and small, from overarching theories to instructions for working through specific kinds of creative blocks. There's so much debate at this point about what constitutes the *true* TRIZ that whatever I say about the system will probably offend some of the TRIZniks.

Following are the nuggets of TRIZ that I believe to be the most significant. Those familiar with TRIZ will notice that I have left out many concepts that are convoluted or difficult to summarize.

Psychological inertia. Altshuller was one of the first people to observe and study the mental blocks that inhibit creativity—what he called "psychological inertia." He observed that our minds are "shackled" by what we already know, and he dedicated himself to examining those jail cells. The creative block was at the center of Altshuller's system. It was the disease he sought to cure.

The contradiction. Altshuller was most fascinated by one particular type of mental blind spot that he called "the contradiction." He noticed that engineers often became convinced that they had to make a tradeoff between two desirable qualities that were "impossible" to achieve at the same time. For instance, imagine that you're building an electric car; you will require a lot of battery power so that the car can travel for hundreds of miles before it has to charge up again. But as soon as you begin adding batteries, you also add weight, making the car so heavy that it won't go very far. The obvious solution (add more batteries) seems to cancel itself out. Faced with this paradox, your imagination may shut down, and you may find you can't seem to generate any ideas. Altshuller observed that it's when we believe that we have to pick between A and B that our mind fails—because we forget to search for a third or fourth or fifth possibility that skirts the seeming conundrum. For instance, what if you could transform the entire car into a giant battery and use its structural parts—like the hood and the door—to store energy? Indeed, right now, several groups of researchers are working on this ingenious work-around; they're developing carbon-fiber material that would transform the car's body into a power-storage system. Patrik Johansson, a physics professor involved in the field, has pointed out that this approach involves a whole new way of thinking about batteries—no longer would the power-storage devices be considered a "burden" on the car. Instead, the batteries would *become* the car (or the laptop, or the phone). It's the kind of imaginative leap—away from the tradeoff and into the impossible—that delighted Altshuller.

"Solve problems that do not yet exist." Gordon Moore constructed an elegantly simple method for forecasting one particular kind of technology (the integrated circuit). Altshuller aimed to find universal patterns of evolution that could be applied to any kind of technology so that he could project trends into the future. He used his science-fiction stories, like "The Donkey Axiom," as laboratories for futuristic thinking. But unfortunately, his scheme for predicting future technologies does not have the pared-down genius of Moore's Law—it's a jumble of rules, theories, trend-watching tools, and so forth. It's unclear whether anyone besides Altshuller himself could effectively put the system into practice, much less prove it works.

The ideal machine is no machine. Altshuller preferred to find a solution that worked as if by magic, with no need for switches, levers, fuel, or human intervention. For instance, what if you wanted to create a window that opens automatically when the temperature in the room goes above eighty degrees? Since metal springs expand when they're hot and contract as they cool down, why not figure out how to use this property to control the window? Altshuller pointed out that too often, engineers come up with complex fixes and bypass the simplest ideas. He advocated a deep knowledge of materials science so that inventors could identify metals and chemicals that could perform tasks that would otherwise require complicated machinery. The right materials could respond to different conditions with a kind of built-in intelligence and even heal themselves.

Solving with swarms. Altshuller also suggested that his students should imagine that the entire machine was made out of a "crowd of miniature dwarfs." These tiny fairies would come together to do a job, like lubricating a gear or creating an electrical current. If you could interview all of those fairies, what would they say? How would they solve the problem? This method anticipated the field of nanotechnology in which swarms of micro-machines come together to perform tasks.

The Ideal Final Result. Altshuller suggested that if you're stumped by a technical problem, you should allow yourself to explore a fantasy land beyond the rules of physics. If you'd like to walk from Berlin to Boston in five minutes, you could shrink Planet Earth to the size of a city block. If you want to deliver water to the top of a mountain, you can command a river to run uphill. And of course you're free to build anything you like out of gold, diamonds, and even the horn of a unicorn. His insight was this: by embracing the impossibly ideal, we can transform the way we see the problem and open ourselves up to new strategies for finding a solution.

For instance, what if you need to paint the inside surface of a U-shaped pipe? Once you think of the word *paint*, your mind will leap to paintbrushes. And from there, it's easy to fall into a trap and assume that you *have to* use brushes to solve the problem.

But what if you owned some enchanted paint that would obey your commands? Then you could yell at the paint: "I order you to fly into that tube and stick to its inner walls!" The paint would leap out of the bucket, wriggle into the tube, and intelligently stick itself exactly where you want it. This fantasy turns out to be a useful one for recasting the problem, because once you've imagined paint flying through the air, you have let go of the idea of brushes. And that will inspire you to imagine all kinds of ways that paint *can be* encouraged to fly. For instance, if the pipe is aligned in the right way, you might be able to plug the bottom end of it, attach a funnel to its mouth, and fill it with paint. In this way, you would use gravity and pressure to "command" the paint to obey your wishes and attach itself to the deep recesses inside the pipe—and then when you unplug the pipe, you would "command" the excess paint to fly away.

I have come to believe that TRIZ is something of a MacGuffin. Altshuller's creation has diverted the attention away from Altshuller himself, from his political struggles (which we will return to later) and his attempts to launch an entirely new intellectual discipline. TRIZ became a corporate darling in the West, while Altshuller himself fell into obscurity—even many people who used his system knew little about his life and writings, and his dream of creating a new field of study. That's a shame, because some of the questions he raised were more intriguing than the answers supplied by his TRIZ system.

In 1956, Altshuller and his friend Rafael Shapiro published a manifesto in the Russian journal *Problems of Psychology.* With that article, they proposed a new *science of invention.* Their article is still almost entirely unknown in the West. Obscure as it is, this article represents the groundbreaking moment when the field of Inventology came into existence. It proposed startling, heretical ideas:

1. Inventive skill can be learned.
2. To study the inventive mind, you must study inventions. "Research on the psychology of inventive creativity cannot be conducted separately from studying the basic laws of technological development," Altshuller and Shapiro wrote.
3. Would-be inventors can learn skills by retracing the breakthroughs that led to existing technologies. "Before operating on living people, a surgeon spends a long time training in the anatomical theater. In the same way, an inventor must systematically analyze earlier inventions. It is also very important to know the history of technology and to be acquainted with every branch of technology in its change and development," Altshuller and Shapiro wrote.

When they published this article, few scholars had tried to understand the workings of creativity and imagination. Even as late as 1950, J. P. Guilford, the president of the American Psychological Association, lamented that his discipline had almost nothing to say about the human mind operating at its peak.

However, by the 1970s and 1980s, a few American psychologists had begun to ask the same kinds of questions Altshuller had raised earlier. It's unlikely that they had been influenced by Altshuller's writings—little (or none) of Altshuller's work had been translated into English at that point in time. Instead, the Western scholars simply happened to become interested in the same mysteries. Because they were psychologists, their questions began with the mind rather than with the technology: What inhibits the imagination? What is a creative block? What kind of lab experiments will tell us something about how people invent?

Only a small cadre of researchers specialized in the study of the inventive imagination, but they turned up some profound results that still resonate today.

Steven Smith, a psychology professor at Texas A&M University, hypothesized that bad ideas can have a far more powerful effect on us than we know. In the 1980s, Smith noticed that engineers and product designers often mistake a clichéd idea for an ingenious breakthrough. When reaching for an original solution, they end up inadvertently copying a familiar concept that has already failed.

Smith and his colleagues called the problem "design fixation" and decided to study it in the lab. Would it be possible to infect engineering students with a bad idea that would actually inhibit their ability to solve a design problem? In order to run the experiment, he needed to find a challenge that the students had never tackled before, and so he picked takeout coffee.

This was in the era before fancy cafés had sprung up on every city block; at the time, when you ordered takeout coffee at a diner, it came in a Styrofoam cup sealed with a flat lid—so that if you took a gulp, you might scald your tongue. "Nowadays, Starbucks and everybody else has solved the problem in a million ways, and we're all familiar with all kinds of inventive coffee cups," he told me. But in the eighties, the to-go cup was deeply in need of improvement, so Smith decided to test the students by asking them to reinvent it.

He divided the engineering students into two groups and described the challenge to them. He showed Group 1 a blueprint of a bad solution—a portable coffee cup that had been outfitted with a straw; Smith told these students *not* to replicate this design, because drinking hot coffee through a straw is idiotic. The second group of students came to the problem fresh, without seeing the blueprint.

Just as Smith suspected, the students in Group 1 couldn't seem to free their minds from the wrong answer; they drew cups that used straws even though they knew that the design would cause people to burn their mouths. Something about that blueprint captivated them; once exposed to it, they found it difficult to think of an original idea. Meanwhile, the students in Group 2 were more likely to come up with novel and effective solutions to the coffee-cup problem.

"If I give people a misleading clue, they'll fixate on that misleading clue. And that makes it much harder for them to find the right answer," Smith explained. "You can even tell them that the information is wrong. It doesn't matter. Even when people know a clue is bad, they can't help trying to use it to find the answer. Once you know something, you can't un-know it."

Smith's study helps to explain why groupthink is so insidious. When we encounter someone else's botched design, we're likely to seize on it. Why? Because our brains like shortcuts. In his brilliant book *Thinking, Fast and Slow,* Daniel Kahneman proposes the "law of least effort"—that is, we always prefer to avoid mental work, and so if an idea comes easily to mind, we feel intuitively that it must be right—even when it's obviously wrong. Because you're wired to take the shortest path to an answer, your brain might seize on an easy question like "How can I improve on this straw?" as a way of avoiding difficult thinking. And for this reason, you might entirely miss the relevant question, which in this case would be as follows: "How can I design a drinking mechanism that reduces the temperature of the coffee by mixing it with air as it leaves the cup?" That second question makes our brains work, and our brains don't like it. So instead we cling to the familiar solution.

Fixedness is so powerful that it can infect an entire organization or industry, with thousands of designers sharing the same blind spot. Smith offers one example of this: In the early years of the railroads, train cars looked like a line of stagecoaches strung together with chains. To cross from one train car to the next, conductors had to leap through the open air, risking a fall or a dangerous encounter with sparks and cinders. It took decades for inventors to come up with a solution to this seemingly simple problem. Why? "When you go from one technology to the next, you carry contextual baggage along with you," Smith said. People who grew up riding in horse-drawn carriages had trouble grasping the possibilities that had suddenly opened up because of the steam engine. When they encountered a railroad car, they saw a horseless carriage. They *didn't yet* see that these carriages could be connected with passageways and vestibules that would make it safe to walk between compartments. No doubt, we're in the grip of similar prejudices today.

In fact, that nineteenth-century term *horseless carriage* reminds me of a phrase that we're throwing around a lot lately: *driverless car*. Most of us grew up thinking of freedom as a steering wheel and a learner's permit. I vividly remember the first time I stomped on a gas pedal and zoomed out onto a highway; the car pounced with such force that I was thrown back in the seat. That sensation is still burned into my mind. For that reason, I will probably always think of autonomous cars as "driverless"—and if you're of a certain age, so will you. To correct that bias, we have to do extra imaginative work in order to open our minds to new possibilities.

An intriguing study conducted by a psychology professor, Michael Mumford, suggests that in some cases the cure might be simple: use your mind to time-travel.

Decades ago, when Mumford was based at Georgia Tech, he created a fake corporate memo that described a nonexistent technology—a "holographic TV." Mumford wanted to see what happened when he asked people to come up with advertisements for the fictional product.

"You've got to bear in mind that this study was conducted in the late 1980s," Mumford told me. Back then, the holographic TV was clearly science fiction; the volunteers in his study—all of them Georgia Tech students—would be well aware that the product did not exist. Nonetheless, they would have to think about how to advertise it as if it did.

Mumford divided the volunteers into two groups. Some were told to start writing scripts for radio and TV ads immediately, without pausing to imagine how the holographic entertainment system would work or who would want to buy it. Meanwhile, the other students were instructed to engage in "mental research" into the problem itself for fifteen minutes; they were encouraged to ask themselves, for instance, why consumers would want to turn their living rooms into a holographic theater. "We told them to imagine the problem in multiple ways and write down multiple definitions of the problem," Mumford recalled. He didn't give them any detailed instructions about *how* they should use the mind's eye; instead, the point was simply to push the students to do some imaginative work. After this period of inner exploration, the students in the second group were asked to write scripts for advertisements.

Afterward, Mumford asked professionals in the ad industry to judge the results. As you might have guessed, the advertising executives were more impressed by the second group's work. The judges were so impressed that they offered advertising jobs to two of the students who had performed the extra mental work. Mumford's results suggest it might not matter *how* you imagine the future but rather that you *do* spend effort imagining it.

Vannevar Bush first began constructing the Memex in his imagination in the early 1930s. It would require a decade to complete his vision of the Memex and publish his description of it. That timeline gives us a hint about what he might have been doing inside his imagination: he may have tried out hundreds of thinking experiments, may have entertained and rejected many concepts, and spent innumerable hours summoning up visual pictures.

Though Bush championed machines that would help us remember and find information, he also believed that the deepest imaginative work could be accomplished only in the wet jungle of neurons. Artists reach "into the unknown with beauty and versatility, erecting on the mundane thought processes a thing of beauty," he wrote. And the artist's imagination—so deeply weird and personal—"will always be barred to the machine."

PART IV

CONNECTING

12

THE GO-BETWEENS

IN 1707, A BRITISH ADMIRAL WITH AN ECCENTRIC NAME—Cloudesley Shovell—was leading a fleet of ships toward home and safety, or so he thought.

Sailors of that era depended on a system called dead reckoning to navigate in the open sea; you would start with your last-known position on the map and then estimate your speed using a clock or hourglass to find your current position. Of course, it was easy to make a mistake, and as the mistakes compounded, ships veered far off course. As for Admiral Shovell, he was so befuddled by the dead reckoning calculations that he pointed the fleet toward the dangerous shoals off the coast of Cornwall. The result was one of the most tragic naval accidents in history: ships crashed and flung sailors into the sea, killing more than two thousand of them. Shovell's body was found in a sandy inlet; according to legend, thieves plucked the bejeweled rings off his fingers. It was the final indignity for the admiral who had been dead wrong.

In the wake of the disaster, members of British Parliament agreed that the country must invest in a new technology that would improve naval navigation. In 1714, Parliament passed the Longitude

Act, promising to award a fortune—£20,000, equal to about £2 million today—to the person who discovered an accurate method to calculate longitude aboard a ship. It was widely assumed that a renowned astronomer would win. After all, England's Royal Observatory had been founded to collect celestial observations for naval navigation.

But instead a key piece of the answer emerged from a surprising source: a carpenter and clockmaker named John Harrison. Because he was a meticulous craftsman, Harrison was able to produce a marine clock that was accurate down to the second. This new technology allowed sailors to calculate longitude by comparing the precise clock readings with the position of the stars and the sun; it was *time*, as much as celestial maps, that turned out to be key to solving the problem.

Harrison's triumph reveals something important about the nature of invention. There are certain people who—by luck, design, or some quirk of personality—are able to bring together knowledge from several fields. They inhabit the cracks and interstices between different disciplines. Harrison, for instance, knew how to craft precision gears and springs; he could descend into a tiny world where moments were measured in slivers of metal. But he was also able to zoom out and connect his knowledge of timekeeping with the art of navigation. His ability to connect won the day.

In Part IV of this book, we will look at how invention benefits from open systems, and specifically from the people who thrive in a connected world. Breakthroughs often happen when we allow unlikely collaborators and odd bedfellows to share our problems, or when we leap across boundaries. But, unfortunately, there are also a lot of impediments to that kind of openness: egos, hierarchies, customs, and the profit motive. In the chapters that follow, we will look at what goes into the making of a John Harrison—a person who can move between realms to stitch together the elusive solution. We'll also ponder what it takes to create a "zone of permission" where new ideas can thrive.

In 2004, Karim Lakhani became fascinated with John Harrison's surprising ability to crack the mystery of longitude. Lakhani, who is now an associate professor at Harvard Business School, thought that Harrison's triumph might offer an important clue to the nature of creativity. And he realized that today's problem-solving competitions could be used as a tool to study the creative mind. "Back in the 1700s, it took an act of Parliament to create a Longitude Prize," he told me. But nowadays, "anybody with a problem and a credit card can in fact go in and conjure up a crowd." Organizations such as NASA, Procter & Gamble, Eli Lilly and Company, Philips, and Cleveland Clinic hospital are all involved in some sort of open invention process, in which they announce their unsolved problems to the world and offer awards for the best answers.

These challenges do more than just generate novel answers to difficult problems; they also create data that we can examine for clues to *how* people figure out the tough problems, and *who* solves them.

InnoCentive, the largest of the problem-solving marketplaces, has been called the "eBay of innovation." It acts as a broker between the companies (or nonprofit groups) and the strangers who might be able to answer their questions. Since its founding in 2001, InnoCentive has built a worldwide community of more than three hundred thousand "solvers," so it draws on a menagerie of chemists, biology students, mechanics, do-gooders, filmmakers, underemployed dreamers, retired engineers, and a guy who calls himself a "fractal expressionist." When InnoCentive posts a challenge, usually a few hundred of these diverse competitors vie for the prize money. Each contestant submits a white paper describing a solution to the problem; a panel of judges picks the best answer; the winner then collects a prize that ranges from a few thousand dollars to a million dollars.

Lakhani recognized that he could mine the results of InnoCentive challenges to find out what differentiates the winners from the losers. "John Harrison was a nobody, and he ended up winning the [Longitude] Prize," Lakhani said. So, if you looked at the evidence, would you find that modern-day John Harrisons consistently beat the esteemed scientists who seemed to be positioned in the "right" discipline?

To answer such questions, Lakhani and his colleague studied data generated by 166 InnoCentive problem-solving competitions. He discovered that indeed the John Harrisons did prevail. Outsiders, rather than those within a field, tended to hit on the best answers. For instance, if InnoCentive posted a question about manufacturing a chemical that is used to make polyester, a person *outside* the textiles industry would be most likely to win. (In fact, an attorney named David Bradin beat out hundreds of competitors in just that kind of chemistry challenge in 2002.)

Lakhani points out that every discipline shares the same library of solutions and also the same blind spots. That's why it can be so powerful to tap the knowledge of someone outside the field—that person will come equipped with a different model of reality, a different set of tools, and his or her own library of solutions.

In 2012, Adam Rivers, a postdoctoral research associate in the marine sciences department at the University of Georgia, was reading the InnoCentive newsletter when a challenge caught his eye. A food company had a problem with a health shake it was trying to develop; the drink suffered from "unwanted food coloration." Reading between the lines, Rivers guessed that the iron supplement in the drink had reacted with some other ingredient to create a disgusting color.

For a food scientist, the problem must have seemed terribly difficult. But for Rivers, the problem seemed easy. He immediately thought about a reaction that occurs naturally in seawater and causes blood-colored stains.

That weekend, he shopped at Walmart for capsules full of green-tea extract; he mixed the powder into a glass of water and then stuffed a piece of steel wool into the mixture. "After a few hours, the glass turned purple," he said. The next step was to figure out how to remove the color without altering the taste or consistency of the shake. It took him another day or two to figure out that trick, and then he submitted his answer to InnoCentive. Rivers beat out nearly two hundred competitors to win a $25,000 prize—a staggering payoff for a weekend spent playing around in a kitchen.

Under ordinary circumstances, of course, a food company would not hire an oceanography student to figure out its health-shake problem. The walls between disciplines and industries keep problems hidden away from the people who possess the oddball knowledge necessary to solve them. This is the advantage of open challenges and contests; they are founded on a sort of productive humility. In order to work with InnoCentive, clients must admit that they don't understand their own problem and that they have no clue who will solve it.

Dwayne Spradlin, former president and CEO of InnoCentive, told me that running the competitions opened his eyes to just how hard it can be for clients to maintain an open mind and accept help from outsiders. One time, he met with some doctors who were trying to solve a conundrum related to immunology; the doctors had considered hosting an online challenge, but they doubted whether random strangers on the Internet had anything to teach them.

"There is no way on Earth that somebody from outside of this field will ever be able to contribute meaningfully to this," one of the doctors had insisted.

Spradlin urged them to try anyway.

The doctors went ahead with their experiment and posted on InnoCentive. Soon, contestants sent in white papers offering solutions. This was a blind-judging process, of course, so the doctors had no idea who had submitted the ideas. Spradlin told me that this blindness discomforted the doctors; they wanted to know the names and pedigrees of the people who had submitted the papers.

"That idea, number 27 right there, I bet you that came from Harvard, didn't it?" one of the doctors told Spradlin.

The answer, Spradlin said, "was no—it actually came from a retired commercial researcher out of India who just wanted to keep his skills strong and had a really cool idea." As Spradlin watched the doctors jump to conclusions, he wondered just how often we miss ideas simply because they come from the "wrong" person.

Lakhani pointed out that our preconceived notions about who should solve a problem often get in the way of actually solving it. "If some kid in Estonia comes up with a brilliant idea, then most people will dismiss it," he said. "Research shows that we have a strong bias against ideas that come from outsiders."

After Lakhani and his colleague released preliminary results of their study of the InnoCentive data, they heard from a professor emeritus at Stanford University. He asked whether Lakhani had looked at the gender of the winners. After all, women had been excluded from science and engineering for centuries. So would this bias against women translate into an advantage when it came to problem solving? If Lakhani's theory was correct—and people who'd been excluded from the inner circle possessed valuable and overlooked ideas—then women should have an advantage in an unbiased competition.

Lakhani decided to dig into the data to find out whether women were more likely than men to win InnoCentive challenges. The results are startling: the women who participated in InnoCentive competitions outperformed the men, even in fields traditionally considered to be "male," like engineering and chemistry. The effect is so strong that if the competitor is female, she is 23.4 percent more likely to win the invention challenge than a male competitor.

Lakhani and his coauthor on the study argue that women represent a huge pool of untapped talent. Even talented women "are, on the whole, more likely to be in the 'the outer circle' of the scientific establishment," they wrote. Because of the bias in fields like engineering, a woman with a good idea may not have the chance to suggest it—unless she competes anonymously in an online challenge. Furthermore, women may have gathered a different set of strategies and proficiencies than their male colleagues—and since these solutions tend to be less well-known, they're more valuable. "Trained and talented individuals who could not enter core positions in their fields, read 'women scientists,' might be more capable of approaching problems in fresh ways," according to Lakhani and his coauthor.

"I don't want to sound like I'm boasting, but this seemed like a very simple problem," Vaishali Agte told me. She had beaten out more than two hundred people to win the InnoCentive challenge called "A Diabetic Cookie That Tastes Like the 'Real Thing.'" Agte is a perfect illustration of Lakhani's theory. A biochemist who has worked for a government-funded research center in India, she knows an enormous amount about how chemicals affect our bodies. But like Adam Rivers, she lacked the kind of connections—the "insider-ness"—that would have made her an obvious candidate to solve a snack-food problem. So instead, she gained entrance through the open door of an online challenge.

When she read about the cookie competition, she felt she had a good chance of winning. The challenge stated that the cookie had to have a glycemic index of about 45—meaning that it would have to use little or none of the sugars that are dangerous to diabetics. At the same time, the cookie had to taste like a bona fide dessert.

To me, this sounded like an impossible problem. But Agte thought it was a snap. As a biochemist specializing in nutrition, she knew what ingredients diabetics could eat. "I made the cookies myself," she said. "Then I asked my friends and colleagues to be my tasters. Based on that, I chose the recipe that was the best according to all of us."

The company that posted the challenge must have discovered a valuable idea in Agte's submission, because it was willing to pay $5,000 for the right to use her recipe.

I was hoping she would share the secret of her cookie with me, but she says that the rules of the competition forbid it. However, she did share her strategy: she waits for just the right problem to pop up on the InnoCentive newsletter and doesn't waste her energies on any challenge that would require a lot of work. Agte points out that if you lose a challenge, you receive no pay at all.

According to Lakhani's research, the winners of InnoCentive challenges pursue exactly that strategy. Most winners (72.5 percent) reported that their idea evolved from a solution they already knew about. Often, they picked up a technique or a concept in one place, and like bees carrying pollen, they'd fly to a new flower. Scanning the list of InnoCentive's top solvers, you notice that a lot of them have acted as Go-Betweens, tying together knowledge from two or more disciplines. They're willing—even eager—to step outside their job titles. One winner, Mounir Errami, told a journalist that he prefers to explore "fields that I haven't mastered." That way, "I am not limited by the known and unknowns or the rules of the field." Though Errami is trained as a biochemist and bioinformatician, he chose to tackle an InnoCentive challenge in cosmetic dentistry.

Warped

The French term *déformation professionnelle* wonderfully sums up the way our jobs shape our minds; the term implies that every profession comes with its own worldview. If you're a structural engineer, you'll glance at a rotten beam in the foundation of a house and see trouble. If you're a microbiologist, the rotten wood seems like a wonderful place to harvest fungi. And if you're a whiskey maker, the smell of the wood might inspire you to infuse your liquor with peaty flavor notes.

Every form of expertise includes a belief system about what is possible or impossible, what is pointless and what is essential.

The winners of problem-solving challenges seem to be relatively free of *déformation professionnelle.* They're interested in connecting, cross-pollinating, and zigzagging. For instance, Adam Rivers—winner of the milk-shake challenge—said, "I enjoy reading broadly and figuring out how to use tools from other fields. There are people who are deep subject matter experts. But that's not me." He said that lately he has become fascinated with software to mine information from social networks, and he's exploring what might happen if he repurposed these tools to answer questions about the ocean. In other words, Rivers is fascinated by what happens if he combines a *this* and a *that.* His urge to forage for combinatory bits and pieces led him to read through the problems posted on the InnoCentive site. "Even before I ever participated in a challenge, I was curious about what kind of problems people had," he said. He realized that the list of challenges could give him clues about the questions outside of oceanography. And he wanted to act as a Go-Between, matching a problem in one area with a solution from a wildly different realm.

Picasso made a surreal trophy out of a bicycle seat and some handlebars—he arranged the two found objects into a bull's head with horns. It is a deceptively simple sculpture—the genius is in the way that he put together two parts. Many inventions, similarly, are born when someone takes to "found" technologies and merges them in a novel way.

Take, for example, Tom Laughlin. In 1997, after being laid off from a job as a project manager at Johnson & Johnson, Laughlin began performing his own experiments. Losing the job "gave me the opportunity to . . . do wild and crazy things," he said, like studying a problem that had fascinated him for years: how to make a fake tan look real.

The hitch was that DHA, the active ingredient in sunless tanning lotions, tended to spread unevenly, and it was almost impossible to control the amount that ended up on your face or back. The lotions left piebald patches all over the body. And if you dabbed on too much cream, you might end up the color of an Oompa Loompa, because it was also hard to control how deeply the chemical permeated your skin.

As Laughlin searched for an answer, his mind turned to the auto industry and a clever method that car companies used to coat metal with a pristine layer of paint. The car would be driven into a special booth and paper sheets would be taped over its windshield and other areas that needed to be kept free of paint. Then spray nozzles filled the air with a mist of paint; these airborne droplets would settle all over the surface of the car, forming an even layer of liquid that would dry into a uniform color. Laughlin decided to build a similar booth—for people.

Seized with this idea, he transformed his Texas house into his laboratory. On his back deck, he created a chamber that looked like a shower stall, its walls pocked with nozzles: he envisioned an auto-painting rig scaled down to human size, one that would coat the body with an even layer of sunless tanning chemicals. But when he'd finished his prototype, Laughlin was afraid to test it on himself. So he wrestled a department-store mannequin into the booth, using it as a kind of crash-test dummy to assess the safety of his person-painting booth. Eventually, he dared to step inside, stand under the nozzles, and press the button. "There was tremendous mist swirling everywhere," he recalled.

In the late 1990s, Laughlin worked with Palm Beach Tan, a chain of salons in Texas, to offer the first spray booths to customers as an alternative to the UV tanning beds. Laughlin also began filing patents for his invention; he describes those patents as unique because they introduced "the concept of spraying a person in a booth" in a manner inspired by the auto industry.

He felt he could make this leap only *after* he lost his job. This was a common theme among the people I interviewed. Even those with successful careers as corporate inventors dreamed of quitting so that they could *really* invent. In the next chapter, we look at how organizations can help or hurt the Go-Betweens—and what might be done to improve matters in the future.

13

ZONES OF PERMISSION

FOR MOST OF HISTORY, INVENTORS WERE DISPERSED IN SHOPS and smithies dotted across the landscape, and they often catered to the needs of small communities. Even as late as the 1870s, families settling on the American prairie would mend their own coffeepots, nail together hog-slaughtering stands, and repair wagon axles. "Every active and ingenious farmer should have a good workshop and his own set of tools for repairing implements," wrote a columnist of the time. Back then, a town was not just a collection of houses but also a gathering place for blacksmiths, tinkers, seamstresses, and cobblers who manufactured the accouterments of daily life. Inventors weren't remote experts; they lived next door.

So it must have seemed strange to people of that era when Thomas Edison built an idea factory in Menlo Park, New Jersey. There, he installed a team of engineers—the "muckers"—who lined up at workbenches to collaborate on overcoming the technical hurdles that would bar the way to light bulbs, batteries, and phonographs. Edison, the mucker-in-chief, tossed out suggestions and checked blueprints, guiding his men as they translated his ideas into wondrous machines. Edison recognized the power of engineered serendipity and constructed his compound at Menlo Park as a vast machine to churn through thousands of trial-and-error experiments to explore the unknown. "I have not failed ten-thousand times . . . I have succeeded in proving that those ten-thousand ways will not work. When I have eliminated the ways that will not work, I will find the way that will work," he once said. The centralized R&D center was born in an era when horse-drawn carriages delivered blocks of ice to kitchens—and yet Edison's concepts would guide the computer age.

By the 1950s, corporations were investing millions in labs and hiring a professional class of engineers, product designers, and scientists. By the late twentieth century, it had become common to assume that new technologies had to be born in a campus occupied by a priesthood of engineers. The Edison model had triumphed.

It was a paradox: an open garden of creativity that flourished behind high walls. To travel to Bell Labs in the 1960s, you drove down a private road and then approached the face of an enormous slab covered in glass that reflected the clouds, as if the building wore mirrored sunglasses. Only certain elite scientists were invited to participate in the meetings that took place behind those mirror shades; so in many ways, Bell Labs was a closed system. However, for those who were invited inside, it was a model of free exchange, a hive of activity where engineers and mathematicians wandered through a playground of oscilloscopes, picture phones, laser beams, and blackboards, riffing on one another's thoughts like free-jazz musicians improvising a tune. Architect Eero Saarinen put a vast atrium at the hub of the central building to encourage chance encounters; he'd envisioned a Brownian motion made of people, where colleagues would collide and pick up each other's energies and then ping back to their workbenches. And indeed, Bell Labs helped to give birth to the transistor, the laser, solar cells, information theory, UNIX, satellite transmission, and on and on.

You might say the nineteenth- and twentieth-century R&D centers—like Bell Labs, Menlo Park, GM's Delco compound, and Xerox PARC—were a response to limitations that no longer exist today. They were a kind of Internet-before-the-Internet, a "zone of permission" where just about anything—or anyone—you needed could be conjured up within a day or two. Supply closets were crammed with rare materials. The libraries housed obscure scientific journals, and in the late 1960s, you could use one of the world's first computerized card catalog systems to speed up your search. The Saarinen building was designed to hold five thousand employees in its wide-open spaces, so this was crowdsourcing that involved a flesh-and-blood crowd.

In Jon Gertner's book *The Idea Factory*, he suggests that the scientists at AT&T had an advantage because of the localized openness that existed within the company. In the 1930s and 1940s, Gertner observes, AT&T's telephone system was a mess, and its "needs were so vast that it was hard to know where to begin explaining them." The cables sizzled in the rain, and the wires were echoing with feedback. Bell Labs was what one of its researchers called a "problem-rich environment," and these "good problems led to good inventions." Because Bell Labs scientists were often the first to learn about the failings of AT&T's infrastructure, they had an advantage over their competitors who were outside the information loop. You might say that AT&T held a monopoly over the problems related to information transmission—the linemen, switchboard operators, and repair crews could gather detailed, rich knowledge about the failure in the system and pass it along to the scientists. AT&T's monopoly therefore gave it unusual advantages in the invention game: not just the deep pockets, but also a deep penetration into the unknown. Long before the Internet, it could gather crowds and conduct its own problem-solving challenges.

But the system that was supposed to engender openness and novelty sometimes produced the opposite result. "For all the undeniable glory of Bell Labs, there emerge little cracks in the resplendent façade of corporatism for the public good," wrote Tim Wu, author of *The Master Switch.* "When the interests of AT&T were at odds with the advancement of knowledge, there was no question as to which good prevailed. And so, interspersed between Bell Labs' public triumphs were its secret discoveries, the skeletons in the imperial closet of AT&T." Its inventors developed magnetic tape, cellular phones, fiber optics, fax machines, and a host of other key technologies. But management scuttled these projects for a variety of reasons—for instance, because bosses worried that the upstart technology might threaten profits from AT&T's landline phones. Or it may be that the managers simply did not recognize the potential of the gizmos that materialized in the labs.

Even the most permissive institution says no sometimes. And this gets at the downside of inventing inside of a closed system, even one as glorious as Bell Labs: an exclusive lab might shut out the unusual and unexpected Go-Betweens—like those we met in the previous chapter—who possess a crucial clue to a solution. And, too, while the campus setting may deliver "good problems" to the inventors, it may also isolate them from their constituents.

Free Radicals

Earlier in this book, I mentioned a study in which researchers identified the corporate engineers and designers whose ideas had led to hit products and millions of dollars in revenues for their companies. You'll remember that Abbie Griffin and her colleagues interviewed these star inventors (whom they called Serial Innovators) in an attempt to discover their formula for success. The researchers found that these inventors often busted out of the confines of the R&D lab so that they could meet with customers and "embed" themselves in farms or hospitals or drugstores. These high performers spent a lot of time connecting dots, and many of those dots lay outside their own company. "If there is something [that the inventors] need to know, they learn it, no matter how far afield it may seem from their original backgrounds and training," according to the researchers.

Managers—often understandably—balked when the inventors pursued a risky and expensive scheme, especially one that pushed the company into unfamiliar territory. But the high-performing inventors refused to back down. They were "willing to risk getting fired for the ideas and products they believe in. Almost half of the [top inventors] in one firm had put their jobs on the line at least once in their careers over important breakthrough product ideas," the researchers wrote.

So the most gifted inventors will combine knowledge from several domains to come up with a novel solution, but the very originality of the idea often makes it a liability inside the company. In fact, the most radical breakthroughs often turn out to be irrelevant to the industries in which they're born—as when a pharmaceutical chemist stumbles across a recipe for cooking up an artificial sweetener. Therefore, the cross-pollinator might have to create his or her own "zone of permission"—either by lobbying and politicking inside a company, by leaving a job and switching into a new field, or by working out a truce with a boss.

Chuck Hull faced this dilemma in the early 1980s. At the time, he worked as an engineer at a company that built ultraviolet lights used in factories to harden a plastic veneer onto tabletops or rubber tiles. The technology allowed you to create a piece of laminate in a customized, two-dimensional shape; you did this by manipulating the way that the UV light hit the surface of the liquid plastic. Hull began to imagine what would happen if he could stack up layers of plastic to form a three-dimensional object; that, he realized, would allow him to sculpt anything he desired. He had stumbled across a path that could transform 1980s lamination equipment into a miraculous new technology that would later be called 3-D printing.

(A point of clarification: In Chapter 7, we followed Tim Anderson and Jim Bredt as they refined 3-D-printing technology in the 1990s; that story took place *after* this one. Tim and Jim were entering the industry pioneered by Chuck Hull.)

One of the keys to Hull's insight about printing objects came from his previous job as a designer of chemistry lab equipment. He knew how difficult it could be for engineers to create bespoke parts like knobs and buttons. To do that, they had to reach out to a professional toolmaker and try to convey their plans for a three-dimensional object in a drawing. The toolmaker would then use the instructions to craft a customized mold, which would then be sent to a parts fabricator. Whenever you made a mistake, you had to start all over again. Hull knew that this was infuriating for engineers, who could spend months perfecting the prototype for one small part. In fact, the difficulty of fabricating customized parts was "killing the US automotive industry at the time," Hull said.

And he'd just found an answer. A 3-D printer would allow engineers to prototype a part on their own, within a day or so, removing a major hassle from their lives.

But when Hull approached his boss and proposed the 3-D printer, he was met with discouragement—after all, the company produced ultraviolet lights, not *Star Trek* replicators. Eventually the two men reached a compromise: Hull would dedicate himself to UV lamps by day, and at night he would be allowed to hang around the company workshop and cobble together his dream machine.

He began by writing the digital code to tell his machine how to cut each layer of plastic and combine the layers into an object. "I was limited to fairly simple shapes," he told me. One day, for instance, he brought home a doll-size cup to show his wife. "It looked like that thing you buy in the drugstore to wash out your eyes," he said. And his first machine "was so kludged together that it looked post-apocalyptic, like some of the equipment they used in that movie *Waterworld.*"

But it worked. So eventually, Hull raised funding and spun off a company to sell the world's first fully realized 3-D printer.

Because the machine was too heavy to lug to demos, Hull shot home movies of his invention and showed these to executives. "The movies were pretty corny," but even so, he said, "we got a tremendous response." Particularly in Detroit. "Back then, the US automotive industry had fallen way behind Japan," Hull said, and the car companies were desperate for a secret weapon. The 3-D printer was just that: engineers could create their own prototypes for door handles or stick-shift knobs, allowing for what is called rapid prototyping, a new method in which engineers could create a "rough draft" of an object and then tweak it, fix it, or rethink it.

Hull's after-hours project launched an industry that is now transforming the way we manufacture goods. And his story illustrates how difficult it can be to introduce a new set of possibilities into the world. He told me that even though engineers embraced the 3-D printer, their managers often balked at spending hundreds of thousands of dollars on a machine designed to churn out rough drafts and improvisations. To them, the 3-D printer looked like a very expensive machine that encouraged sloppiness; they had trouble grasping that it would revolutionize their industry. So Hull had to imagine his way across boundaries, and then figure out how to make his breakthrough relevant to strangers in an entirely different field.

Hull's story reveals just how difficult it can be for inventors to nurture the most original ideas, the ones that give birth to new industries. He had to create his own zone of permission in the after-hours at his workplace, when he became his own boss. Because he was connecting insights from several industries—plastic lamination, auto design, and lab equipment engineering—he had to find a kind of no man's land where it was possible to nurture a technology that at first seemed to fit nowhere but would eventually spread everywhere. The development of 3-D printing offers a lesson in the fertility of open labs, hackerspaces, garages, and wizard's dens. Most of the breakthroughs in the 3-D printing field emerged from people working independently. You'll remember that Building 20—that glorious wreck—opened up a space where a semi-employed artist and a grad student could build a desktop 3-D printer out of trash.

Meanwhile, S. Scott Crump hit on another key process used in the 3-D-printing industry in 1988, while trying to mold a customized toy frog for his daughter. Working in his kitchen, he filled a glue gun with polyethylene and melted wax in order to sculpt the toy layer by layer; soon, his wife commanded him to move the experiments to their garage. Like many other legendary garages, this one became the birthplace of a company that would pioneer a new industry: Stratasys helped to open up a consumer market for 3-D printing.

There may be no greater zone of permission than a garage, a dorm room, or your office at midnight, after your colleagues have left for the day. Facebook started in a dorm room, Yahoo! in a trailer on the Stanford campus, and Google in a garage. The first inklings of Twitter began with a scrawled piece of paper in an Oakland apartment. Often, the idea flourishes at the fringes of a university, where the inventor can benefit from the intellectual hubbub without being subjected to rules and plans.

In 2008, two public-policy researchers asked, "Where do innovations come from?" To answer that question, the researchers tracked the winners of the R&D 100 Awards, which *R&D Magazine* hands out to spotlight excellence in invention. They found that in 2006, only six of those one hundred awards went to Fortune 500 companies. The authors of the study hypothesized that large firms now focus on existing products instead of hunting for radical breakthroughs. As a result, "many talented scientists and engineers have voted with their feet and have left work in corporate labs in favor of work at government labs, university labs, or smaller firms"—the kinds of places that tend to win the R&D awards.

A 2009 survey confirmed these findings: the survey found that inventors who work in academic labs or tiny companies (fewer than a hundred employees) tend to generate a disproportionate share of the most valuable ideas.

But if you're now imagining this highly productive inventor as a teenage Zuckerberg in a dorm room littered with pizza boxes, you might want to revise your mental image. The 2009 survey also reveals that US inventors hit their peak at forty-seven years old, on average. Chuck Hull conforms to this trend: he was in his mid-forties when he began the experiments that led him to pioneer a new industry. Perhaps middle-aged inventors are so productive because they enjoy more independence than their younger peers. With gray hair come the seniority, sabbaticals, a home workshop, and the savings account that allow for the pursuit of risky projects.

Of course, now that the tools of invention are becoming cheap and ubiquitous, it's possible for many more people to throw themselves into building the "impossible" or "useless" machines that end up transforming our world. This accessibility promises to be a game changer for invention. We now share a vast public atrium where ideas ricochet among billions of people.

The Future of R&D?

When I interviewed Martin Cooper, originator of the cell phone, he pointed out that the way we communicate is intimately connected with the way we discover and develop new opportunities. "The most important thing . . . cell phones did in society is make us all more productive," he said. Now that we're all connected "with the ability to collaborate almost in real time," inventors are free to gravitate to the projects that fascinate them rather than having to wait for orders. "It reminds me of that quotation: 'If you want to get people to think out of the box, do not create the box in the first place.' And yet, the corporation is founded on boxes," Cooper said. This is why he predicts that the hierarchical, centralized, industrial-age R&D is destined to die out.

After all, it's just about impossible for any private effort to compete with the twenty-four-hour, worldwide hackathon on the Internet. Lakhani, the Harvard Business School professor, told me that increasingly "creativity is becoming a numbers game" because a million people on the Internet can almost always outsmart a team of ten engineers. "It's profound that the tools for creation are becoming democratized," he added.

Of course, many miraculous products and profound ideas continue to emerge from the corporate campus; Google, Facebook, and Apple are proof of that. And yet, even these modern-day R&D giants are finding ways to tear down their own walls and interact with the crowd. Apple's app store, for instance, opened up a creative marketplace where independent developers could sell their wares. In so doing, Apple benefited from the ingenuity of thousands of people who reimagined the iPhone as a light saber, a stethoscope, a UFO detector, a dog whistle, and a blood-testing device.

Indeed, key information is almost certain to be hiding in what economist Eric von Hippel calls the "dark matter," among the users who possess the richest knowledge of our shared problems. They're the doctors, athletes, housecleaners, plant biologists, car mechanics, and farmers whose daily experience exposes them to opportunities that are invisible to everyone else.

For that reason, many organizations are figuring out new ways to open up their own R&D efforts and collaborate with the inventors out in the dark—and to leverage the power of the Internet.

Doing so requires enormous creativity on the part of managers, who are reinventing themselves as talent scouts, ethnographers, cool hunters, crowd wranglers, and data miners. Indeed, to fully master this new style of R&D, organizations have to mirror the Internet itself.

Houston, We Have a Problem Solver

Hila Lifshitz-Assaf, a business-world ethnographer, had a chance to witness exactly that kind of transition at NASA. While earning her postdoctoral degree at Harvard Business School, she "embedded" herself at the space agency, where she attended meetings, read memos, and interviewed dozens of people, watching as engineers and managers struggled to redefine themselves in the age of the Internet and open problem solving. She wanted to find out how the scientists would react when they had to share "their" work with outsiders.

Lifshitz-Assaf observed that one turning point in 2009 seemed to have caused a good deal of angst. For decades, NASA had been searching for a reliable way to predict when the sun will throw off high-energy particles; this kind of foreknowledge is essential, because the solar particles can harm astronauts and equipment during a mission. NASA scientists had not been able to devise a method that would allow them to forecast solar events with the accuracy that they needed. Could the crowd do it?

To find out, NASA offered a prize of $30,000 to whoever submitted the best method for predicting solar-particle storms. More than five hundred people competed for the prize, and a retired engineer living in rural New Hampshire won. Using his own equipment, Bruce Cragin devised a new technique that could forecast the onset and duration of solar activity with about 75 percent accuracy—an enormous improvement over the existing method. Here was proof that the crowd could solve an "impossible" problem within just a few months.

The leadership at NASA was "stunned" by the way the crowd

had outperformed its own people, according to Lifshitz-Assaf. Some NASA engineers and managers felt humiliated. She interviewed several of them who badmouthed the new system: to them, it felt as if the agency was parading its problems around in public and asking for help—which implied that its own engineers had failed. "'The professional shame was palpable," wrote Lifshitz-Assaf. One of her sources at NASA told her, "It's extremely frustrating. The feeling . . . is now 'What value am I?'"

But other managers and scientists flourished under the new system, even finding new and imaginative ways to prod the Internet into supplying solutions. For instance, one NASA engineer realized that she could be most effective as a talent scout. When she needed a specialized medical device that would operate in space, she searched online and discovered that a small-town doctor, in his garage, had hacked together exactly the device that NASA required. She reached out to the doctor, asked to see a prototype, and discovered that it worked well enough to be used aboard the International Space Station.

Other engineers also recognized that they'd be far more effective if they became skilled at drawing upon minds *outside* the organization. "The invent-it-ourselves model is generally not sustainable" these days, one source at NASA told Lifshitz-Assaf.

Organizations like NASA are asking new questions about R&D and invention. How best to match the problems and solutions? In some cases, it may make sense to move the engineers and designers out into the community, where they can observe problems firsthand, rather than isolating them on campus. It also makes sense to encourage the people who understand problems—doctors, social workers, teachers, oil-rig workers, truckers, and so on—to participate in the R&D system. In the next chapter, we look at one case study to understand where R&D might be headed.

14

HOLISTIC INVENTION

If you've spent time in hospital, you've had a chance to observe both wondrous technology and worrisome lapses. Here, surgeons deploy robots to repair arteries and nurses program pumps to deliver pulses of medication through the night. And yet even amid these miraculous machines, it's clear that something has gone terribly wrong. In 2013, the *Journal of Patient Safety* reported that each year one out of every six deaths in America—more than four hundred thousand people—was the result of a preventable failure that happened inside a hospital. Some of those errors, no doubt, resulted from the poor design of medical devices.

In Part I of this book, we saw that Lead Users are often the first to diagnose problems. After working with a tool for hundreds of hours, they become intimately familiar with its flaws. But unfortunately, when they create a fix for themselves, user-inventors often don't have the time or skills to disseminate it. That's a particular quandary in the health-care system, where doctors and nurses rarely collaborate with engineers and entrepreneurs who could turn their knowledge into lifesaving improvements. Ideally, institutions would find a way to connect the problem finders to an R&D pipeline.

Since 2000, the Cleveland Clinic has been striving to do just that. Insights generated by the hospital's idea-harvesting and innovation programs have led to more than twenty-seven hundred patent applications and seven hundred approved patents. Cleveland Clinic has launched more than seventy spinoff companies to develop medical products; its doctors have dreamed up the concepts behind devices like an artificial heart that weighs less than a pound and can keep patients alive while they're waiting for a transplant. The program gives us a glimpse of what R&D looks like when an institution draws from knowledge that ordinarily wouldn't be captured. The most unusual feature of Cleveland Clinic's program might be its bracing honesty; staff are encouraged to point out failures, to ask, "Why does it have to be this way?" and to challenge the habits of the institution itself.

In this chapter, we will follow the story of one Cleveland Clinic doctor in order to examine the potential—and the difficulties—of this new method.

In 2012, Dr. Yogen Saunthararajah, an oncologist and hematologist, was worrying about the central-line catheter. His patients shower, eat, work, and sleep with a catheter burrowed under their skin, sometimes for months. The tubelike device offers an efficient way to deliver chemotherapy treatments and other medications into the bloodstream. But when it's infected with bacteria, the catheter can transform into a killer, spraying poison into the blood. Saunthararajah and his staff often had no way of knowing that a patient had gone into septic shock until they noticed his or her fever, and by then it might be too late. Saunthararajah explained that saving the patient "is a devil-and-deep-blue-sea type of situation. You have to take the infected tube out. But at the same time, you need some way to pump the patient full of antibiotics and fluid to keep them alive. So while you're taking out one tube, someone else has to start putting a new one in."

Hospitals depend on the central-line catheter to deliver drugs to patients in intensive care, pediatrics, cardiology, and chemotherapy wards; and nearly three million Americans a year will be outfitted with one of these devices. Though the catheter can be a lifesaver, it has a vicious downside. About thirty thousand Americans die every year from infections caused by the central-line catheter, making the problem almost as deadly as breast cancer.

In theory, there is a way to prevent a catheter from causing an infection: clean it. Groups like the American Society of Anesthesiologists have released checklists and guidelines to make sure that nurses scrub the catheters with antiseptic. Studies have shown that rigorous hygiene can dramatically reduce infections.

But it may be that by focusing attention solely on hygiene, we have failed to detect all of the ways in which the catheter kills. A 2012 study found that after nine days inside the body, the tubes grow ever more dangerous—even when you do your best to clean them. And the fact is, many parts of the device *can't* be cleaned because they're inside a vein. "You can only get to the part that hangs outside the body, which is called the hub," Catherine Musemeche, a pediatric surgeon, told me.

When a catheter becomes infected, the failure is often blamed on the carelessness of nurses or poor hygiene in the hospital. Sometimes that blame is well deserved. But all the finger-pointing at nurses and facilities may have drawn attention away from the catheter itself. Indeed, on Saunthararajah's ward, the medical staff had tried to find new methods to clean the devices and had stepped up their hygiene protocols, with only limited success.

One morning in 2012, Saunthararajah was trying to relax after a night spent agonizing about a patient who had been infected by a tube. As he read the *New York Times*, a headline caught his eye: "Teeth That Think." The article touted a low-cost device that could sniff out danger inside the mouth, describing it as "a superthin tooth sensor (a kind of temporary tattoo) that sends an alert when it detects bacteria associated with plaque buildup, cavities or infection." Sauntharajah marveled at this clever concept and began imagining how a similar strategy could be used to tackle the catheter problem. The idea—taken from dentistry—"gave me a window into another world," he said. In the oncology ward where he worked, everyone assumed that bacterial infections *are, by their very nature, invisible to the nursing staff.* But now he realized that it didn't have to be that way. With the right tools, maybe his nurses could detect dangerous bacteria before they spread or caused harm.

Seeing a possible answer transformed the way he felt about the tubes stacked up in his supply closet. He and his nurses had focused all their attention on scrubbing the devices, carefully wiping the exit site with antiseptics. He realized that approach had blinded him to another way of seeing the problem. Now he began to wonder whether the devices could be built so that they would never infect the patient.

He asked himself why we're stuck with a "stupid" catheter in the era of smartphones. "I have an iPhone in my pocket that lets me search through all medical journals. Technology has moved ahead by leaps, and yet here I am inserting into patients almost the same kind of tube that we've been using for fifty years," he said.

He began to imagine what you might call the Ideal Tube—one that would include a built-in sensor as an early-warning system. The tube would take readings inside the body, sniff out dangerous bacteria, and send messages to the medical team—perhaps the tube could issue a text message or sound an alarm at the nurses' station. That way, the staff could catch a problem long before the bacteria had a chance to grow.

Of course, Sauntharajah didn't know how to build the device that he imagined. And he might have forgotten about his insight and moved on, if not for the invitation he'd received from his colleague at the hospital, Dr. Christine Moravec. She's a research scientist who manages one of the innovation projects at Cleveland Clinic, and she routinely sends out e-mails to the hospital staff asking them to notice problems and propose inventions.

"My Hippocratic Oath is to make people better. That's my professional motive," Saunthararajah told me. But "it's hard enough for a specialist like me to keep up with my own team, let alone to dabble in bioengineering." Saunthararajah wouldn't have been able to tackle inventing a piece of equipment on his own, so he was pleased that the hospital offered somewhere to go with his ideas.

When he first approached Moravec, she wasn't sure whether the central-line catheter *needed* to be transformed into a smart device equipped with a biosensor. But after she researched the scope of the problem and learned that the central-line tube is one of the most dangerous pieces of equipment used in the hospital, she realized "that the solution to this could be really important."

Though Saunthararajah had suggested the concept, building a prototype would require several more sub-inventions. How could you create a bacteria sensor small enough to fit in a tube, one that could sample the bloodstream without itself causing infections? And how would the sensors inside the tube send a warning to doctors and nurses?

When faced with difficult questions like these, Cleveland Clinic sometimes turns to the crowd of online problem solvers. Moravec told me that the hospital posts challenges on InnoCentive because this method can be cheaper and faster than hiring in-house engineers. "If we have a question we cannot address here, can we get four hundred thousand people throughout the world to work on it?" she said. "We posted six challenges during the first year [2011] we worked with InnoCentive, and we found good answers to five of those challenges," Moravec explained. "People might say, 'Are you kidding me? Someone comes up with a solution and you give them $20,000 and then you turn it into a commercial device and make a lot of money off of it?'" But, she said, "the reality is that Cleveland Clinic has to invest millions of dollars into testing the product. We have to spend quite a lot of time and effort testing the solution."

To pursue Dr. Sauntharajah's idea, the hospital posted a challenge on InnoCentive offering a prize of $20,000 to "design a biosensor or 'early warning system' that reports when a central venous catheter has started to become contaminated." The hospital received 394 answers, and Dr. Sauntharajah worked with a bioengineer to sort through them and pick the best one. (Because of intellectual-property agreements, Sauntharajah couldn't tell me exactly how the winner conceived of the technology inside the smart tube.) The next step will be to create a prototype and then to partner with a company to develop the product and perform safety tests. Translating the plan into an off-the-shelf solution will no doubt involve a long process full of setbacks and financial risks. It's still not clear how the story will turn out, because the path from initial concept to finished product takes many years. But already this new idea has helped to start a conversation about how we might invent our way around some of the terrible infections that begin in hospitals.

In 2014, a team of Spanish researchers published a paper in which they announced that they had designed a prototype of their own smart central-line catheter. It is equipped with sensors that "know" when a film of bacteria is forming on its surface. And, just as Sauntharajah had envisioned, it can send a warning to medical staff.

Health-care breakthroughs have often started with medical MacGyvers who figured out why patients were dying and then hacked together inventions that solved the problem. Take catheters themselves: many key developments emerged from the minds and hands of physicians. For instance, in the 1960s a medical student named Thomas Fogarty figured out how to combine a urethral catheter and a tiny balloon to create a device that could inflate inside the patient's artery. This let him catch a blood clot and drag it out through a small incision. The Fogarty catheter has been called the first minimally invasive surgical device and has saved millions of lives. But, "I couldn't get any manufacturer to make the catheters," he told a reporter later. "Companies thought I was some stooge fooling around. I didn't have any credibility." The device ended up being commercialized only after Fogarty's friend brought it to the attention of an electrical engineer. If not for some lucky turns, Fogarty's breakthrough might have remained a quirky piece of homemade equipment that never spread beyond his own practice. It makes you wonder: What intellectual wealth have we lost because doctors lacked the right connections? Or because the hospital discouraged medical staff from speaking up about a problem?

Often a promising idea never even gets off the ground because of the walls between disciplines. The doctor who watches patients die doesn't know how to reach out to engineers who could help to design the fix that would prevent those deaths. "One can think of a major academic hospital as a 'problem-rich environment' and an engineering university as a 'solution-rich environment,'" wrote Dr. Stephen Schimpff, the former CEO of the University of Maryland Medical Center. The trouble, he points out, is that "physicians are in their hospitals and engineers are in their own setting, and the two rarely meet."

Moravec asserts that openness is the future of biomedical research—and that other top hospitals, like Harvard Medical School, have also embraced online problem solving. But zeroing in on the flaws in the system—and admitting you don't have the answer—can be challenging for any institution. To post a question on InnoCentive, "you have to make your weakness obvious to the world. You have to be willing to say, 'We don't know how to do this.' Nobody wants to be vulnerable like that. But we realized that we have to be willing to take that risk to move forward."

It is no wonder that so many organizations thwart the ideas that bubble up from below. Inventors are the cousins of whistleblowers—they point out flaws in the system and flaunt mistakes. We tend to think of invention as a politically neutral activity, but it is not. Watchdog groups, critics, and reformers are some of the most powerful drivers of technological improvement.

The Watchdogs of Design

Photographs from the 1930s give us a shocking glimpse into the carnage that was commonplace on American roads. What today might be a minor accident—like hitting a curb—could be a death sentence back then. With little structural support or stability, the auto would flip, crumple, cave in, or even wrap around a tree as if it were made of rubber. By the time passengers were brought into emergency rooms—decapitated, crushed, and mangled—it was often too late to do anything for them.

After witnessing gruesome casualties, some physicians were so outraged that they lobbied car companies to provide safer vehicles outfitted with restraints to prevent passengers from flying through the windshield. A few doctors became so concerned that they rigged up seat-belt systems for their own family cars, and crash pads for the front passenger seat, which was then nicknamed the "death seat."

Unfortunately, it would take decades of avoidable deaths, and the publication of Ralph Nader's *Unsafe at Any Speed* in 1965, before seat-belt laws passed in most states. In addition to arguing for safety restraints, Nader's book documented tragic engineering failures, like axles in the Chevy Corvair that could buckle and collapse as the car rounded a corner. "Nearly one-half of all the automobiles on the road today will eventually be involved in an injury-producing accident," he wrote. Nader argued that automotive engineers received almost no education in human-factors design and labored in environments where they could not advocate for lifesaving improvements. "The liberation of the engineering imagination for automotive safety cannot take place within the automobile industry," he contended. Nader's point was that the designers and executives at the car companies had begun to dwell inside their own parallel reality—one that encouraged them to ignore their obligations to the customer.

Nader illustrated this argument with a chilling story about a banker who wrote to General Motors in 1954, describing how he'd braked suddenly and his eight-year-old boy, riding in the passenger seat, flew forward and broke his tooth on the dashboard. The banker requested that GM add padding to its dashboards.

GM replied to the banker's suggestion with a bizarre letter from safety engineer Howard Gandelot. The GM engineer suggested that instead of worrying about the design of the car, parents should take responsibility for their children's behavior. "As soon as the youngsters get large enough to be able to see out when standing up, that's what they want to do," Gandelot wrote. He said that he'd trained his own boys to grab the dashboard when they stood up in the front seat—his idea being that the children's arms would suffice as a safety feature. When his sons forgot to hold tight, Gandelot confided, he'd pump the brakes, so that the boys lurched forward. That would teach them.

Ralph Nader, though he is not considered an inventor, may have done more to reimagine the cars of the 1960s than anyone inside the auto industry. *Unsafe at Any Speed* revealed engineering failures and missed opportunities to save lives, and it led to consumer outcry that radically reshaped the American car.

Inventing, at its best, can be a form of civic engagement. When we notice a problem in the designed environment, we have an obligation to speak out and participate in improving it. At the same time, we all know how challenging it can be to step up in that way; pushing for change usually requires facing off against powerful interests that have a stake in the status quo. Many of us—quite reasonably—hesitate to raise our voices when we notice a problem. Yet, this kind of "civic invention" can be the most rewarding of creative acts.

In the next chapter, we will return to Genrich Altshuller, the Soviet dissident who championed the courageous imagination of the reformer.

PART V

EMPOWERMENT

15

PAPER EYES

AFTER A STRONG CUP OF COFFEE, YOU MIGHT FIND YOURSELF entertaining a new possibility; in a feverish flush, you pour out notes, or scribble sketches, and an audacious idea begins to take shape in your mind. But after a few days, the doubts set in. A chorus of naysayers howls in your head: Who do you think you are? If this notion is so brilliant, why hasn't someone tried it before? Is the solution possible, or just a mirage?

Of course it's crucial to entertain doubts and listen to feedback. But too often, we shrink away from a novel idea because we're afraid or overwhelmed. In fact, fortitude might be the most important ingredient in any inventive effort—without it, an idea remains just a thought or a scribble.

In 1931, a patent inspector named Joseph Rossman published the results of an unusual survey. He had reached out to hundreds of inventors and asked them about their mental life and their creative methods. His survey included this question: "What are the characteristics of a successful inventor?" There was a surprising concurrence among the responses. About 70 percent of them said "perseverance."

This final part of the book will investigate a fifth challenge to the imagination: How is it that we feel empowered to solve a problem, and how do we persevere? The next two chapters will explore the kind of inner strength that is necessary to champion original thoughts.

We met Genrich Altshuller earlier in this book, and I want to return to his story now, because he pioneered a psychological theory that wonderfully illuminates the courage it can take to imagine. He endured torture, imprisonment, hard labor, and starvation in pursuit of free expression. To understand his hard-won insights, we must travel back in time, to the Soviet Union of the Stalin era.

When Genrich Altshuller was a teenager in Azerbaijan in the 1930s, he roamed the fantastical worlds of Jules Verne, disappearing into the pages of books to explore the ocean floor or fly to the moon. He received his first patent—at the age of about fourteen—for diving gear.

While he devoured science-fiction novels and tinkered as a boy, he seems to have been oblivious to the dangers of these pursuits. Stalin was then conducting a campaign of terror, what became known as the Great Purge, and many heroic people would die defending their ideas. For instance, Nikolai Ivanovich Vavilov—one of Russia's most brilliant agricultural scientists and founder of the world's premier seed bank—advocated crossbreeding to create plants that would yield a secure food system. Vavilov had collected high-yield varieties of wheat and potatoes, and though he was not an inventor in the conventional sense, his discoveries had the potential to revolutionize Soviet agriculture and end its terrible famines. But Stalin regarded Vavilov as a dangerous subversive. Thus, the man who might have delivered millions of people from hunger died of starvation in a Soviet prison cell.

It's not clear how much Altshuller knew about Stalin's Great Purge while it was unfolding during his childhood. But as a young man, when he worked in a patent office, Altshuller had the chance to witness the creative darkness that had fallen over his country. He was dismayed by the terrible quality of the patent applications stacked up in the office. It hadn't yet occurred to him *why* so many of Russia's inventors were inept. Instead, he became fascinated with the failures. What was going wrong in people's minds, and how could he fix it? So began his intense period of research in the late 1940s, when he read through thousands of patents to discover principles that he believed would lead anyone toward an ingenious solution.

During those years when he spent long hours in the patent library, he stumbled across evidence that suggested that—contrary to state propaganda—the Soviet patent rate had fallen almost to zero. It finally began to dawn on him why this was so: Scientists had been shot, starved, tortured, and disappeared. Their books had been pulled from the shelves, and their names had been expunged from the records. "The leaders of the Union of Inventors had been arrested and crushed," Altshuller said later. "Nobody was left." Stalin had appointed his own idea men and celebrated their achievements, but then they too disappeared. "How could we talk about inventing?" Altshuller lamented. And yet he did.

In 1948, Altshuller and his best friend, Rafael Shapiro, penned an open letter to Stalin in which they argued that creative problem solving could be taught in Soviet schools and factories. It was daring and reckless to instruct Stalin on any matter, but Altshuller believed he could get away with it. He was a brash young man, with the lanky physique and the chiseled profile of a film star; and now he was convinced that he wielded a mental sword that gave him the power to cut through any problem.

Altshuller and Shapiro sent their letter to Stalin, *Pravda*, and a dozen or so government ministries, with tragically predictable results: the military police ambushed the young men, arrested them, and charged them with a long list of political crimes. Altshuller ended up in a prison where he was subjected to horrors that became a crucible for his ideas. In the midst of his tribulations, his mind tilted in a new direction—and he dreamed a bold and weird science-fiction vision that predicted the moment we find ourselves in now.

Today, we tend to assume that invention is *apolitical*, a wealth generator that can be skimmed by companies and governments. But invention can also topple governments and empires. It is inherently dangerous and difficult to control.

The imagination can become our own free and wild territory. When all else fails, it can offer a refuge. This was clear to Altshuller in October 1950, as the police dragged him to a prison much like the one where Vavilov had died seven years earlier. Soon after his arrival at Lefortovo Prison, Altshuller was hustled by guards through the dank passageways and slippery stone stairs to the interrogation room and shoved into a chair in front of Captain Malyshev, a bureaucrat with the eyes of a dead fish.

"Well, have you decided . . . to confess?" Malyshev asked, sounding bored.

Malyshev wanted names. His job was to convince the prisoner to accuse his friends. But Altshuller refused to speak, and the torture began. Malyshev ordered the prisoner to stay awake until he was ready to sign a confession.

Altshuller was not even allowed to close his eyes, and if he lay down on the bunk in his cell, guards would rush in and beat him. Soon, he felt so muddled and depressed that he knew he was about to crack. Another night of this and he would spew names; he would implicate his dearest friends and family. He had to come up with a solution.

That's when it occurred to him that he still had his magic power: he could invent. Sleepy as he was, he began concentrating on how to build a device that would protect him from torture. He'd learned during those years of reading through thousands of patents that he had to frame the question in just the right way. Only then could he hit upon the simplest and best solution. So he started by carefully stating the design challenge to himself. He had to be able to sleep throughout the day, in order to prepare himself for the all-night interrogation sessions; but simultaneously, the guards who peeped into his cell had to *believe* that he was awake. And how to accomplish that? Altshuller realized that he could solve the problem by redesigning his own eyes so that they could be simultaneously open and closed. To pull that off, he'd need a second pair of eyes, and he would have to make those eyes out of whatever he could find in the cell.

He started with a pack of cigarettes, peeling the paper away from the tobacco and then carefully tearing the paper into two ovals. Then, with the help of his cellmate, Zasetskii, he hunted around the floor and found burned matches. He and Zasetskii used these to draw black dots—just the size of irises—on the scraps of paper. Using saliva, Zasetskii stuck them onto Altshuller's lids. The cell was outfitted with two bunks, one of which was catty-corner from the door and harder to see from the peephole. Altshuller arranged himself on this bunk, leaning against the wall, half propped up on the bedroll. He and Zasetskii began a loud conversation. Altshuller closed his eyes and—bliss!—he blacked out into a profound sleep that lasted eleven hours. While he was out, Zasetskii kept up a one-sided conversation, occasionally adjusting his friend's arms and legs. Meanwhile, Altshuller stared at the world with his paper eyes.

At about ten that night, Altshuller woke up and readied himself for interrogation. He stashed his extra pair of eyes, and when the guards came to take him to the interrogation room, Altshuller swaggered behind them, cracking jokes as if he'd just breezed in from a night at the Bolshoi Theatre.

His torturer, Malyshev, was baffled. By his reckoning, the prisoner should stagger into the room, close to madness and ready to sign the confession. Malyshev demanded to know whether Altshuller had slept during the day.

Altshuller pointed out that this would have been impossible.

Malyshev ordered the guards to step up their surveillance, peeping into the cell more often to check that the prisoner was awake. His invention worked beautifully: the guards failed to catch on to the trick.

One night during his interrogation, Altshuller grew so cocky that he decided to have some fun. He told Malyshev that he was a freak of nature, the one-in-a-million man who didn't need to sleep at all. Altshuller claimed he could last for months without any shuteye. So go ahead with the silly torture! As he ridiculed his interrogator, Altshuller "felt that the power of human reason was something weighty," more powerful even than prison, or the Soviet system itself.

But he was only half right.

During the fifth or sixth day of this torture, Altshuller slept propped up on a bunk, with his paper eyes glaring out at the cell. At some point, though, he fell into such a deep slumber that his chin hit his chest and his mouth fell open. Just then, a guard opened up the peephole and spied what appeared to be a gruesome scene: a dead body on the cot, with its eyes bugged out and tongue hanging out.

Altshuller woke with a start and found himself standing on his feet surrounded by guards—his friend Zasetskii had yanked him upright and simultaneously managed to swallow the paper "eyes." With the evidence gone, the guards never caught on to the exact mechanism of the trick, but they knew Altshuller had cheated.

"If you sleep with your eyes open again, I'll put you in the lockup," the warden said. Altshuller would no longer be allowed to sit on the cot; he'd have to spend the day walking around his cell.

The game was over. So he came up with a desperate Plan B. That night during interrogation, he would have to punch Captain Malyshev. That way he'd be reclassified as a dangerous prisoner and sent off to the lockup ward—where he would be allowed to sleep and he would also be exempted from interrogation for seventy-two hours.

Zasetskii warned him that this was a bad solution: "In lockup, there's no ventilation. It's damp. Hellishly cold."

Altshuller didn't care. "I heard only one thing," he wrote later. "They wouldn't call me to interrogations and would let me sleep six hours a day. What more does a simple Soviet prisoner need?"

That night, the guards dragged him to the interrogation room and dropped him into the chair by the door. For two hours, Altshuller planned out his attack on the captain. He was so sleep deprived that his thoughts crawled along and he couldn't figure out how to throw the punch.

And then he noticed a pitcher full of water on the table. In his muddled state, it seemed to him that the only way he could bear to hit Malyshev was with this weapon rather than his own fist. Slowly, as in a dream, he lurched to his feet and toward the pitcher, reaching out for it.

Malyshev jumped up, reprimanding the prisoner for taking a drink of water without permission. They grabbed the pitcher at the same time and wrestled over it.

At that moment, the door flew open and three men marched in, including a Soviet colonel. In the dim light, the colonel loomed as large as a tank, rolling toward them.

Altshuller let go of the pitcher and heard it crash on the floor, and water sprayed everywhere. Suddenly, he felt wide awake.

The colonel barked at him, "Why are you resisting the investigation?"

"It's not being conducted properly," Altshuller replied, and then launched into a speech about the sleep deprivation that violated his rights as a citizen. Of course, the colonel would have known all about the torture and possibly even designed it himself. But that night, the colonel turned his fury on his underling.

"You don't let him sleep?" the colonel yelled at the interrogator. And Malyshev trembled, confused because he'd only been following orders.

The colonel turned to Altshuller and said, "Catch up on your sleep, and then we'll have a talk." He sent the prisoner back to his cell, where Altshuller was allowed to stretch out on his bunk for as long as he liked. Or at least, that's how Altshuller remembered the story decades later.

I must admit that this final incident in Altshuller's account seems to defy logic. Why would the infamous torturer have acted like an indulgent hotelier? Altshuller himself realized that this part of the story did not seem credible. He suspected that his memory had been distorted by trauma. More than forty years later, when he told this story to a tape recorder and played it back, he realized that he must have repressed whatever really happened that night with the chief torturer of Lefortovo Prison. To survive, he had forgotten the worst and remembered "only what gave me strength to struggle," he theorized. At the same time, his account did capture a certain truth; at age twenty-four, he had felt superhuman, his mind like "a weapon far superior to automatic pistols," for he possessed "the secrets of solving creative problems." He "adamantly believed in the power of reason, its possibilities." This confidence in his own mind—as much as those paper eyes—had allowed him to endure the torture and emerge from Lefortovo Prison without ever betraying his friends or signing a confession.

The punishment for refusing to talk would be harsh. A judge sentenced Altshuller to twenty-five years in the Gulag. He was sent to Vorkuta camp in a coal-mining town above the Arctic Circle, where prisoners entered through a gate that declared "Labor in the USSR is a matter of honor, glory, valor and heroism." The sign echoed the wrought-iron motto of Dachau: "Work will set you free." Here, Altshuller slept in a frigid cabin crowded with men and spent days pushing coal carts and shoveling rocks into horse-drawn carriages. At one point during his imprisonment a supervisor sent him to work with the gravediggers, a team of men who pulled a wagon full of corpses along the railroad tracks that led to the edge of town. When the tracks began to hum and a train chuffed toward them, the gravediggers had to throw their wagon off the tracks and leap out of the way. Dead bodies would scatter, and the train would plow over them. When Altshuller's boots crunched on shards and fragments on the ground near the railroad tracks, he realized that he was stepping on human bones.

"Before prison," he later told a friend, "I struggled with simple human doubts. If my ideas were so important, why weren't they recognized?" But now it seemed to him that his very life depended on his mind. In the labor camp, he couldn't study patents or solve engineering problems; nor could he work on his science of invention. But he could imagine starship captains and undersea worlds. At night, in the crowded cabin, he mesmerized the other men with stories of a far-distant future, narrating space operas that lasted for hours. And so in the Siberian wasteland of frozen mud flecked with bone, Altshuller discovered his gift. He would eventually become one of Russia's leading science-fiction authors.

In 1953, news of Stalin's death filtered into the camp. In the political shift that followed, millions of captives returned home. Altshuller became one of them in 1954, when authorities reopened his case and pardoned him for his putative crimes. He had been radicalized by his time in the Gulag, and he now began to conceive of invention in a new way—as a form of democratic thought and activism.

In the 1950s, most psychologists regarded invention as an unknowable process. They believed that technological insights were wired into the minds of those with inborn talent. Altshuller, on the other hand, assumed that invention could be studied, perfected, sped up, and improved. The human imagination could be *engineered*, just like a machine.

I suspect that if Altshuller had published his manifesto in English-language journals, his idea might have taken off and inspired a new kind of study of the human mind. In the United States, psychologists were beginning to devise "creativity IQ tests" predicated on the assumption that inventive ability was innate. It had occurred to almost no one at that time that ingenuity could be learned or that valuable insights about the human mind could be gleaned from the patent system. "There was strong resistance" to the idea, Altshuller wrote later, about the reaction of his contemporaries in the Soviet Union. "A science of invention . . . threatened more than a few sacred cows. It denied the uniqueness of history's great inventors and intruded upon the common perception of the incomprehensible nature of the creative process."

Altshuller also proposed a system of education in which students would learn to be inventors by studying patents (as he had) and by honing their minds with inventive puzzles—for instance, by figuring out how to build feedback into a gas tank, so that the driver would know when fuel was running low. In 1971, he founded a school like no other in the world, called the Azerbaijan Public Institute of Inventive Creativity, open to anyone who showed up. He pushed his students to solve problems related to refrigeration, agriculture, and factory equipment. But the true topic of inquiry was their own minds. As he threw engineering puzzles at the students, he pushed them to observe how their imagination failed, and to deploy new strategies to enlarge their mental powers. For instance, in one class, he proposed a problem that involved what he called a physical contradiction: You have designed a machine in which a metallic ball drops through a tube, opening and closing circuits as it passes through. But here's the hitch: if the ball fits snugly against the walls of the tube—so that it can close an electrical circuit—it will get stuck. How do you engineer a ball that will conduct electricity *and* move through the tube? His students were stumped. So Altshuller asked them to jump to their feet and enact a dramatic performance. Five people held hands, pretending to be a metallic ball. The others played the part of a tube, forming two lines across the classroom. As the "ball" of students forced their way through the gauntlet, two people in the "ball" lost hold of their companions and broke off from the group.

This exercise illustrated that if you made the ball out of a flexible material—say, liquid mercury—you could achieve the solution: parts of the ball could break off and stick to the walls even as the rest of it fell and continued to conduct electricity. "We have to imagine that an object is made from small dwarfs, small animals, or a swarm of flies or just a cloud," Altshuller told his students. "This gives [us the] ability to feel that the shape is not rigid. . . . It will behave as we need it [to]."

His school was an unlikely portmanteau, a blend of bohemian improvisation and gearhead thinking in the idealistic spirit of Nikola Tesla and Buckminster Fuller. And, alas, it was doomed.

In 1974, a Soviet committee began policing Altshuller's activities and accused him of disobeying orders. In protest, Altshuller left his position at the school and shut it down.

After that, Altshuller traveled from city to city, holding public workshops open to all would-be inventors. He became a kind of Johnny Appleseed of problem solving, spreading his ethos to thousands of students, many of whom set up their own study groups.

"The higher the percentage of creative individuals in the population, then the better and higher that society will be," he wrote. Of course, he allowed, some people channel creativity to evil ends. But for the most part, if people are encouraged to imagine a better way, they will become passionately involved in making improvements. "I saw this in the prison camp," he wrote. Inventing comforted him in the worst of times; when the torturers tried to strip him of his humanity, he held on to his sense of self by retreating into his imagination. He once told his friend Victor Fey that only three kinds of people survive in the Gulag: the pious, the well connected, and the "crazy inventors." Now, decades later, he said that he was teaching his students to develop a "strong mind" or "independent thinking." He dreamed of a society peppered with inventors who were so independent that they would refuse to bend to dictators.

In the 1970s, he set out to reach millions of children by publishing a regular "inventor's page" in the newspaper *Pionerskaya Pravda*. In his column, he asked young readers to figure out problems related to propeller design, candy making, and industrial pumps. A spinoff TV show pitted kids against professional engineers to solve the kinds of problems that arose in factories.

It was as if he wanted to give everyone a new way of seeing—a set of eyes like those he had made for himself in prison. He advocated for an open debate about design that would penetrate down into the tiniest details. The food we eat, the streets we walk, the beds that cradle us—these, he believed, should be designed *by* us rather than *for* us.

He was a kindred spirit to West Coast visionaries—like Myron Stolaroff, Doug Engelbart, and Stewart Brand—in his quest to unite technology with personal liberation. And it's easy to imagine an alternative reality in which Altshuller became a hero to the 1970s hackers of San Francisco who pushed for "computer liberation"—cheap and accessible technologies to boost the power of the human mind. Many of Altshuller's slogans ("The best machine is no machine," "Think for yourself," "Solve problems that don't exist yet") seem like they should have been rendered in trippy lettering in *The Whole Earth Catalog*. But that's not what happened.

Altshuller never seemed interested in making money off his system, and as Russia opened up to the West in the 1990s, he was suffering from Parkinson's disease, which limited his mobility. But many of his disciples emigrated to the United States, Israel, Japan, South Korea, and the United Kingdom, and they brought TRIZ with them. They called themselves TRIZniks and set up clubs and schools, and some of them designed software packages that would help engineers search for valuable ideas in the patent system. Other TRIZniks found jobs as creativity consultants at corporations. And still others wrote dense books on how to use TRIZ in engineering, illustrated with flow charts and peppered with jargon.

Dozens of companies adopted some form of TRIZ training, from Amoco to Dow Chemical to Ford to Samsung. Though Altshuller championed the "unreasonable" imagination of the visionaries, it turned out his methods were also useful for helping Fortune 500 engineers to figure out how to shave production costs or make tweaks to existing products. "With TRIZ, individuals can generate amazingly creative solutions without threatening the stability of the company," according to Subir Chowdhury, a management consultant and author.

This is, of course, almost the opposite of the radical, dangerous, and even "pointless" creative leaps that Altshuller sought to encourage. He once wrote, "Don Quixote's lunge at the windmill is one of the most significant fights in the history of mankind." As he saw it, the process of invention was a path to something even greater: the development of an independent mind.

Nowadays, his educational theories are ripe for rediscovery. In the twenty-first century, hackerspaces and open labs have made invention accessible to millions of people around the world. Thousands of elementary, middle, and high schools now embrace some form of training in R&D, invention, tinkering, or civic engagement through design. It was as if Altshuller's science-fiction imagination had flung him deep into the future and he'd seen the inevitable: that factories would shrink, proliferate, find their way into schools, and then merge with the child's imagination.

16

TINKERING WITH EDUCATION

In the late 1990s, an MIT professor named Neil Gershenfeld decided to offer a class called "How to Make (Almost) Anything" where students would have free run of the R&D lab. He'd intended to train future engineers, so he was surprised when his class was mobbed by students who planned to be artists and architects.

"All my life I've been waiting to take a class like this," one student told Gershenfeld. "I'll do anything to get in," another said. Many of them had fantasized about objects that didn't exist. Now, using the R&D tools in Gershenfeld's class, the students transformed their visions into inventions—from a private booth that recorded screams to an alarm clock that gamboled around the room.

Back then, a suite of R&D tools could cost $50,000 or more, and it seemed like wild speculation to imagine that this technology would ever have wide cultural impact. But already, Gershenfeld was predicting that a machine he called a Personal Fabricator would transform our relationship with stuff—he described this imaginary device as something like the *Star Trek* replicator that ordinary people could run at home. In 2003, when I first heard about his dream of a personal-fabrication future, I bookmarked it as one of those interesting stories I would get around to investigating later, when the technology actually existed.

And then in 2004, I learned that an MIT grad student (and a former teaching assistant in Gershenfeld's class) had already built a briefcase-size factory that could print out eyeglass lenses customized to the needs of each wearer. At that time, as many as a billion people had no access to an optician's services or eyeglasses, meaning that millions of those people could not work or study because they were effectively blind. By creating this portable factory, Saul Griffith hoped to give poor people the power to manufacture their own glasses. Karl Marx once observed that whoever owns the means of production also can control the levers of social conditions and the economy. And now, Griffith had devised a machine with curiously political overtones—the means of production shrunk into a suitcase? I had to see it.

In May of that year, I waited outside a dingy apartment building in Cambridge, Massachusetts. I'd punched the buzzer a couple of times, and finally I poked at the intercom and yelled into the speaker: "Hey, Saul? I thought we had an interview scheduled for today?" After a few minutes more, the front door of the building clicked; I pushed through it and wandered the hallways, until I found Saul Griffith waiting for me in his doorway, a strapping guy with an explosion of frizzy copper hair and a bandage dangling from one hand.

After the initial pleasantries, I pointed to his hand and asked, "What happened to you?" His bandage was beginning to unwind, and underneath it I could see a bloody gash across his palm.

Kite-surfing accident, he told me.

Then he shuffled over to the stove and began frying eggs, gingerly holding the spatula in his injured hand. While he ate breakfast, I bombarded him with questions about how he had come up with the idea for his tiny factory.

Griffith told me that a few years before, he had traveled to Guyana as a volunteer with a nonprofit group that gathered castoff eyeglasses from the United States. In a small village, he had worked in a clinic with a retired optometrist, trying to match glasses to the hundreds of people who came through the door every day.

"It was terribly depressing," Griffith told me, especially when the only pair of glasses that corrected the vision of a macho man turned out to be a pair of pink cat-eye frames. Griffith realized that this do-gooder solution was deeply flawed; it cost about $100 to collect and refurbish each pair of the used glasses in America and deliver them to developing nations. At that price point, you couldn't serve many people.

So he began turning the problem around in his mind. Why was it so hard to deliver customized lenses to the rural poor in a country like Guyana? He realized that when a region lacks infrastructure—like shipping services, roads that reach small towns, abundant electricity, computer networks, and so forth—it's almost impossible to match a client with an item that has been manufactured specially for him or her in a faraway factory. Once he'd diagnosed the problem, he began thinking in a new way: if the Guyanese villagers were effectively cut off from the lens factories, why not bring the factories to them?

As we saw earlier in this book, some of the most gifted inventors take the time to travel, immerse themselves in problems, listen to constituents, and develop a kinship with sufferers. They see the problem through other eyes. Perhaps most important, they feel the urgency to help others—a kind of secondhand pain that cracks open the mind. So Griffith returned to the MIT Media Lab and threw himself into hacking together the machine that he'd envisioned.

When I visited him that day in 2004, he pulled it out of his closet and opened it up: at its center, a piece of silver foil stretched over a circular frame. Griffith had bought the windshield-tinting foil at an auto-parts store and then used it as if it were half of a balloon, inflating or deflating it to achieve an infinite variety of lens shapes.

He only had to pour liquid plastic into the curved mold to produce a customized lens.

His invention had caused something of a sensation in the nonprofit world. "People call you up and say, 'Oh, we love it,' and waste hours telling you how wonderful your idea is," Griffith said. The trouble was he had nothing to sell them; he had made only this one prototype. It would cost half a million to five million dollars to get his product to market, and it seemed impossible to raise that kind of money. And since his many customers made less than $2 a day, he couldn't figure out how to come up with a sustainable way to fund his technology in the long run.

In fact, by the time I met him, Griffith had become disillusioned with his lens-printing machine. Even though his invention had attracted much attention (and a Lemelson-MIT National Collegiate Student Prize), he had come to the conclusion that it wouldn't produce lenses cheaply enough to transform the economics of vision care. He'd created a brilliant prototype but then realized that it wouldn't help the people he'd wanted to serve. "It turned out that we were solving the wrong problem," according to Griffith. Chinese factories could mass-produce eyeglass lenses at rock-bottom prices—so there was little benefit in reinventing the manufacturing system. The challenge was not to make the lenses but to build a public health system that would bring eye-care services (like the diagnosis of vision problems) to everyone.

So by 2004, Griffith was widening his view from that one small machine to contemplate the big picture. Instead of simply delivering a solution to a place like Guyana, could you enable the local people to implement their own ideas? He had become fascinated with the question of how to distribute to everyone the kinds of engineering skills that he'd learned at MIT. You might say that he had now become concerned with questions of Inventology, though, of course, he didn't use that word. The word he used was *education.*

That day when I visited him, he was still struggling to explain his philosophy. He said he needed to show me something and began pawing through a cardboard box. "This!" he said, as he disgorged a musty book with a cracked spine called *The Boy Mechanic,* first issued by *Popular Mechanics* in 1913. He opened it up to his favorite illustration, a picture of a kid who had strapped himself into a homemade glider cobbled out of sticks and canvas. Dotted lines indicated that the boy was about to jump off a cliff and sail through the air and over a river to land adroitly beside a house. There were no adults in the picture. "They're telling you to build a plane and you're twelve years old," Griffith said with delight. His own parents presented him with a how-to-invent book like this one when he was a boy, and it inspired him to become an inventor. Kids need to figure out how to nail together their own wings—but just as important, they should learn how to jump off a cliff. The young inventor must dare, dissent, question, and leap.

As it happened, I'd caught Griffith when he was starting to articulate what would become one of his biggest questions: How can we transform educational systems so that kids around the world can become technological explorers? "Adults are a lost cause. You've got to get the kids," he said.

He pointed out that the first generation of kids who grew up with computers—Bill Gates, for one—learned to be software wizards and hackers. They understood the potential of digital code as their elders never could. Griffith believed that when kids get hold of milling machines and laser cutters and 3-D printers, a similar magic would happen. The next generation would be hacking not just bits of information but also atoms. All those objects that had once seemed so hard to modify or rethink—houses, airplanes, streets—would suddenly look like software. They would be programmable. Griffith and his collaborators (entrepreneur Joost Bonsen and artist Nick Dragotta) had begun to produce comic books for kids that taught inventive thinking. Called *Howtoons,* the cartoons (distributed on the Web and on paper) showed children how they could build a marshmallow shooter, an air-hockey game, a skateboard equipped with ice-skating blades, and dozens of other toys. The materials that the kids used would be available for cheap or free—and often featured trash-picked items like soda bottles. In the same way that comic books can help kids to learn French or Spanish, *Howtoons* was intended to teach engineering skills. Griffith had also created a "Howtoons Clubhouse" in a back room at MIT where he hosted pizza parties for young inventors.

I sat in on one of the parties, curious to see how he'd teach five- and six-year-olds to understand the principles of materials science, physics, and nanotechnology.

Griffith started the proceedings by leaping around at the front of the room and waving his arms.

"So what kind of machine do you want to invent tonight?" he called out.

A few months before when he asked this question, a gang of kids decided to build a hovercraft out of a Toro 220 leaf blower and a wood plank, with a tarpaulin skirt that trapped the air underneath. It turned out to be powerful enough to let the kids—and even Griffith—levitate and fly around the room.

Now, a squirmy boy raised a hand to propose a project. "Spirits," he said. "We could make animal spirits."

Griffith laughed, admitted that might not be possible, and then guided the kids to a table that he'd set up with bowls, chocolate frosting, paintbrushes, and boxes of cereal. He instructed two girls, each about six years old, to dab chocolate onto Cheerios. They then dropped their chocolate cereal, along with regular Cheerios, into a bowl of milk. The girls watched, transfixed, while the Cheerios floated around and seemed to find one another, sticking together in clumps—the plain attached to plain, and the chocolate stuck to chocolate. (The behavior is based on a simple principle: oil and water don't mix.) Griffith explained that if you paint tiny chocolate stripes on your Cheerios you can "program" them to form complicated patterns—like snowflakes—in your cereal bowl.

In fact, he was trying to teach the underlying principles of nanotechnology to these kids. Engineers create nanoparticles that self-assemble into a machine, and likewise, those Cheerios "know" how to move around, find partners, and create a pattern. And so Griffith's cereal lesson was intended to train kids to imagine a microscopic world of nano-things and to think like engineers.

Twenty-First-Century Shop Class

When I interviewed inventors, I often asked them about their early development. Many of them told me the same story about their childhoods: they had been given the run of a home workshop where they played with rocket fuel, sharp blades, or electrical wires. Often, something exploded. Their parents encouraged this dangerous hobby—and often a father or uncle mentored the child. The inventors seldom mentioned learning their craft in school.

Indeed, if anything, school had been an annoying distraction from the hands-on projects that obsessed them. A PhD student in mechanical engineering named Ben Trettel told me that when he was in high school in the mid-2000s, the teachers gave him the impression that shop class "was for kids who were going to end up as auto mechanics." As one of the "smart kids," he was pushed toward AP classes—even those that didn't interest him.

Meanwhile, Trettel developed some of his most useful skills and passions in a home workshop, where his father helped him to build water guns and potato cannons. "I had access to a machine shop basically my entire life," he said, adding that his childhood hobby later helped him to sail through advanced classes in fluid mechanics, one of the most challenging areas of physics. "Other students were confused, but I had already experimented with this stuff when I was fifteen," he told me. "Because of the water guns, I started reading scientific papers about nozzle design. How do you design a nozzle so that a stream of water shoots farther? This is a question at the cutting edge of science. And yet a teenager can work on this problem himself, do some experiments, and actually make some progress with it. And that's kind of cool."

Compelling new research confirms Trettel's experience. It turns out that children who can visualize and manipulate objects in the mind's eye tend to become top achievers in midlife. The psychology researcher David Lubinski has called spatial-thinking ability a "sleeping giant"—a long-ignored talent that is connected with great accomplishments in science, math, engineering, and creative pursuits. "Spatial ability is a powerful systematic source of individual differences that has been neglected in complex learning and work settings; it has also been neglected in modeling the development of expertise and creative accomplishments," he wrote.

Lubinski and his colleagues contacted more than five hundred people who had—as thirteen-year-olds back in the 1970s—performed exceptionally well on tests of spatial reasoning and visual thinking. In 2013, the researchers reported that these young visual thinkers had grown up to become gifted creators of science and technology, who filed patents and published in top research journals. The researchers found that spatial-reasoning tests could be used as a predictor of creative achievement.

Earlier in this book, we saw that many inventors depend on an "inner R&D lab" where they prototype and test out machines. Think of Tesla, who wrote, "It is absolutely immaterial to me whether I run my turbine in thought or test it in my shop." We've also learned that it requires more than just innate skill to manipulate an object in your mind; you have to expend a lot of effort, or willpower, to keep the movie rolling in your imagination. As Altshuller suggested with his term *strong thinking*, the muscles of our imagination can be trained and expanded; with practice comes power. A 2013 meta-analysis of cognitive-science studies helps to confirm what Altshuller suggested: spatial reasoning can be learned and improved.

Everyone R&D

Thousands of educators are now championing hands-on projects and school R&D labs for all levels of learners. The new educational philosophy goes under a plethora of names: "twenty-first-century shop class," "project-based learning," "Fab Lab," and "tinkering school" among them. More than five thousand high schools, universities, and libraries now offer students advanced prototyping tools (like 3-D printers). Meanwhile, organizations including the Department of Agriculture, DARPA, the Smithsonian, the National Science Foundation, and the Department of Education are underwriting programs that encourage children to engineer and invent.

Of course, even as adults redesign education, kids themselves are often several steps ahead. At age thirteen, after his mentor died of pancreatic cancer, Jack Andraka decided that he had to create an early-detection test for the disease. He had spent his childhood tinkering in his basement and competing in science fairs, and so he had developed the audacious belief that he could solve just about any problem, including the diagnosis of a deadly cancer. In 2010, when he decided to take on pancreatic cancer, Andraka spent nights and weekends tunneling deep into the Internet. In online databases, he considered eight thousand proteins that might have potential as biomarkers and sorted through them all.

He performed this initial research for free—leveraging the billions of dollars' worth of basic science available on the Internet. He discovered that some of the journal articles he needed were behind pay walls, and he had to ask his parents to shell out for the fees. He drew upon resources that most kids don't have—like the basement workshop, the ready cash for projects, and the many adults who encouraged, mentored, drove, coached, and nurtured him.

And yet, it's surprising just how much he could do in his room alone with the Internet. He trained himself to read cancer-biology papers, found a promising protein, and then learned about the carbon-nanotube technology that he believed could be used to create a blood test. Again, using the Internet, Andraka contacted about two hundred scientists to ask for lab space where he could run trials to test his ideas; via e-mail he secured a corner spot in a lab at Johns Hopkins.

In 2012, when Andraka was fifteen, his obsessive effort paid off: he devised a low-cost method for detecting a protein in the blood. The hope is that this device will save lives by giving early warning when cancer cells proliferate. This feat catapulted him to fame; he gave a Ted Talk, won an Intel ISEF Gordon E. Moore Award for his invention, and has been acclaimed as the "prodigy of pancreatic cancer." However, his device has yet to be tested in clinical trials, so it's not clear whether it will prove to be an effective screening tool. Whatever the outcome, his story gives an illuminating glimpse into just how far a teenager can go when the entry point is a laptop. "My mind," he wrote, "felt like a powerful weapon I could set loose on any problem."

CONCLUSION

A FEW YEARS AGO, I SAT IN A CAFÉ READING A SHEAF OF ESSAYS by science-fiction writer William Gibson. An editor at the *New York Times* had just sent the galleys to me, and I was savoring Gibson's language, which enters the bloodstream like a drug, producing a mild hallucinogenic effect. When I glanced up from the page and surveyed the city outside the window, I felt as if I were wearing Gibson-glasses. A truck rumbled down the street, farting out blasts of diesel smoke, and I realized that these were clouds of history—the smell and soot of a city one hundred years ago. Meanwhile, at the table next to me, a tattooed kid punched code into his laptop.

Under the spell of Gibson's book, I suddenly understood my surroundings not as a discrete contemporary tableau but as a hodgepodge of the years 1650, 1910, 1980, 2011, 2020, and 2050. "The future is already here. It's just not evenly distributed yet," Gibson once quipped.

In 1990, Gibson and Bruce Sterling published a novel called *The Difference Engine,* an alternative history that takes the uneven-future idea to an extreme. In the novel, the computer revolution happens in Disraeli's era, and the Victorians work out their calculations on steam-powered thinking machines. The book introduced a vision of "steampunk" to a broader audience and also anticipated a fashion movement whose enthusiasts mix corsets with goggles and pearl-handled cell phones. But steampunk is more than mere fantasy. It's all around us. In many cities, the petticoats of Victorian buildings brush up against Wi-Fi hot spots, and if you want to time-travel, all you have to do is walk down the street.

In one of his essays, Gibson describes Tokyo as a city built out of "successive layers of Tomorrowlands, older ones showing through when the newer ones start to peel." Lurking in the back corner of a noodle stall one day, he watched a man playing with his phone. The gadget was glossy, "complexly curvilinear, totally ephemeral-looking," shining with reflections of the city around it. Gibson zoomed in on an accessory hanging from the phone—a "rosarylike anti-cancer charm." According to Japanese pop-culture lore, such talismans are supposed to protect against microwaves.

Even a single object is a conglomeration of many ideas, cultures, and historical periods. The toothbrush originated in China during the Middle Ages; there, people figured out that hog bristles could be affixed to bone handles. Our modern toothbrushes, though they're made out of new materials, look startlingly similar to the ones that people devised hundreds of years ago. If you scan the environment where you're sitting right now, you will discover that most of the technologies that you depend on were developed centuries ago—from the laces on your shoes to the ceramic mug that holds your coffee. We live in a steampunk world, undergirded with hundred-year-old sewage pipes twisted side by side with the fiber-optic cables, all of the tunnels and pipes running through Tomorrowland like the squiggly script of a language that we can't read. Everywhere you turn, you will find a record of people long ago—most of them now forgotten—who explored the impossible, annexing a new territory in the designed environment.

The uneven future that Gibson observed is often stamped with a record of injustice: for instance, hundreds of millions of people still lack access to basic vision care, while others wear glasses implanted with microchips that allow them to tunnel through data as they stroll the sidewalks. Even in the United States, a country fat with wealth, people are suffering because they're on the wrong side of the uneven future.

Dr. Saunthararajah, the hematologist and oncologist at Cleveland Clinic, was troubled by these kinds of disparities. He wondered how it had come to this: at the same time he carried a computer in his pocket, the R&D system had failed to address basic problems that were killing his patients. "It doesn't seem right," he said. "Can't we do better?"

That question kept playing in my head as I worked on this book. We desperately need a robust R&D system that can develop vaccines in time to halt the pandemics, that can wean us off of fossil fuels and provide all of us with nutritious food. This is the maddening, frightening, and bizarre side of life in the twenty-first century. Though we possess the brain power, the talent, and the tools to solve our most worrying problems, it's enormously difficult to organize ourselves around the big questions.

"Amazing possibility on the one hand and frustrating inaction on the other—that is the yin and yang of modern science. Invention generates ever more gizmos and gadgets, but imagination is not providing clues to solving the scientific puzzles that threaten our very existence," wrote Roberta Ness, former dean of the University of Texas School of Public Health. So how to throw ourselves at the big problems? One way might be to engage far more people to detect the dangers around us and then imagine new possibilities. Each of us comes to the problems from a unique vantage point, and we're all fluent in our own language of solutions. So if we can harness the enormous diversity of seven billion minds, we stand a far greater chance of discovering the elegant solutions that lie somewhere out in the unknown.

Many Mothers of Invention

We're the species that invents. Our brains have evolved for problem solving, planning for future disasters, and subsisting in environments like the Arctic, where by all rights we should not be able to dwell. Even our bodies have adapted to our role as inventors. Our early ancestors walked for days and hauled prey on their backs, and their skeletons speak to us of their ceaseless labor—those early humans had strong bones and joints adapted to groaning under heavy loads. But then, twelve thousand years ago, at the beginning of what is known as the Holocene Era, the human body went through a subtle transformation; our leg bones became fragile and our joints became weaker than they had been thousands of years before. "People were adopting farming, domesticating animals. That reduction in physical activity is what's resulted in this light skeleton," according to anthropologist Habiba Chirchir. If you ever doubt your pedigree as an inventor, simply glance down at your knees—those vulnerable joints belong to a species that learned how to coax beasts of burden and machines into doing its work.

Of course, much of our genius arose from our geographical diversity; in settlements around the world, people learned to do these things in different ways and put to use whatever materials were abundant. It is our strength as a species that we speak a babble of languages and come equipped with multifarious memories, passions, and skills. The way people invented has always been tangled up in the circumstances in which they find themselves.

This book is a testament to just how powerful our situations can be in shaping our ideas. To achieve the most valuable kind of breakthrough, you often need to be in the right place at the right time, performing the odd or unusual activity that allows you to open a door that is closed to everyone else. A tennis coach picks up a thousand balls. A pilot tires of lugging his suitcase across acres of airport. A

NASA engineer happens to be playing with a nozzle. An oceanographer stumbles across a milk-shake mystery. These inventors lucked into a situation that gave them unique access to a problem, a discovery, a glimpse of the future, or a bridge between two fields.

And this is why opening up the doors to a diversity of people will transform not just the power of our ideas but the very way that we invent. The tools of manufacturing and discovery have been extended to billions of people who would have been excluded even a few decades ago. The collective might of the human imagination has grown enormously in the twenty-first century. Every dollar spent on basic science stretches much further, because so many millions of people have access to the results and can make use of them. We're also finding ingenious ways of repurposing our knowledge, for instance by storing it in vast databases that can themselves be used to generate new discoveries. Atul Butte—the data scientist and biomedical entrepreneur—rhapsodized about a new era in which biotech tools are becoming available to anyone, and a kid's science-fair project could lead to a medical breakthrough. "That's what democratization really does," he said.

Each of us has experienced something unique, the seed of an invention. You possess privileged information about a hidden problem or a valuable solution because at some point in your life, you have glimpsed a connection that no one else has seen in the same way. For instance, when Dick Belanger was immersed in both glue-gun design and toddler care in the 1980s, his mind swizzled these experiences together and came up with the sippy cup. Back then, of course, Belanger had to bet his own bank account to go forward. But today, inventors like Belanger have many more options for funding their projects, and they can print out prototypes in a matter of hours. Truly, we are at the dawn of a golden age for problem solving.

Several people I interviewed for this book expressed wonder at the speed at which all the barriers are falling. Emblematic of that shift, a mechanical engineer named Tully Gehan has started up a company called Factory For All, based in Shenzhen, China, to act as a sort of concierge for inventors. Gehan "lubricate[s] the process of manufacturing" for people who don't know how to work with factories.

When Gehan first arrived in China in 2005, "it was really hard to make a prototype," he said. "I could go out on the street and there was maybe some lady with a cart and a whole bunch of capacitors, but if I wanted a specific capacitor I'd have to order it from Singapore or even America. It was so painful just to do anything here in China." Now, he said, you can order a custom-made circuit board in under twenty-four hours; you can hunt down whatever parts you need on Alibaba or find them for sale—and immediate delivery—somewhere on the Internet. "And almost any kind of sensor you can think of, somebody has written all of the software for it and debugged it and then blogged about it. That lets lots of people Frankenstein-build a project that used to only be doable by a competent electrical engineer," he said. He has noticed an influx of "English majors" and other non-engineers into the world of designing and manufacturing products.

According to Gehan, "Now twenty-year-olds are raising money on Kickstarter and going to a company and saying, 'What can you do for me?' This is the moment when an individual is able to create a vision and take it into the market. In that sense, we can all be Steve Jobs now," he argued, because "manufacturing can be done on a credit card."

Chris Hawker, the designer from Chapter 4 who helped his client raise $13 million on Kickstarter, believes that we're stepping into a new era of product development. "My job used to be so hard" in the 1990s, he said. Back when he was working on one of his first inventions, an algae scraper for fish tanks, he spent days in the library rooting through patent records and researching manufacturers, scrawling down phone numbers, and then driving home to call up factories. And in those days, if he dared to go forward with a project, he had to spend many years and hundreds of thousands of dollars to find out whether anyone actually wanted to buy the product.

But now, he said, "I can just get on Alibaba, . . . register online, and talk to people in minutes and look up Wikipedia articles" to research and source component parts. He has outfitted his studio with three 3-D printers, and "the product is much more refined because we can iterate it and style it perfectly, and for a fraction of the cost." Once he had to work the phone to find retailers to sell the product, but now they seek him out, because their talent scouts have spotted one of his products in development. When he wanted to launch his Carbon Flyer toy in 2014, he and his team "didn't have to go out and hunt down investors; we don't have to take on any debt," he said. Instead, Hawker posted his prototype on Indiegogo—and within months, more than four thousand backers contributed to his campaign and the money rolled in.

Of course, the centralized, Edison-style R&D system continues to be crucial to the way we keep billions of people fed, clothed, and warm. But we are already beginning to see the enormous potential of the transparent, decentralized, open, and chaotic R&D system that belongs to no one and lives everywhere. My hope is that it will evolve into something like a global immune system—that we can reinvent invention itself and put the designed environment in the hands of the many rather than the few.

Nature uses multiple strategies rather than a monoculture to solve problems. The pathogens that threaten us constantly mutate, changing their tactics, making discoveries, hitting on breakthroughs, finding ways to disarm our drugs. Nature's research and development is based on billions of experiments. It is a resilient, decentralized, and wildly diverse scheme. Cancer cells defeat us simply with their numbers and the trial-and-error ingenuity with which they evolve. How to fight back? By recruiting millions of people to try billions of experiments and thus mount a random sweep through the unknown that might unlock the secrets of killer diseases. This system may also be our best hope for surviving the problems that will be unleashed by climate change—from water shortages to mass extinctions to agricultural disasters.

Apple and Google create terrific products that we love. But corporations are not designed to solve social and environmental problems. We need to evolve another system of invention that mimics the natural world, so that we become a more robust species, better able to survive and adapt. We need an R&D system like the system that protects our own bodies—open, obstreperous, and resilient. It should gain strength from every attack on it. And this system must be intimately connected to the pain receptors rather than cut off from them. The people who suffer from the failures of technology—the Lead Users, the whistleblowers, the sick, the disadvantaged, the poor—are the ones who are best able to diagnose our problems. They are—or should be—at the center of R&D.

ACKNOWLEDGMENTS

MY MOST PROFOUND THANKS GO TO HUGO LINDGREN — FORmer executive editor of the *New York Times Magazine*—who hired me to write the weekly "Who Made That?" column in 2012. Hugo's fascination with the secret history of ordinary objects ignited my own curiosity, and he taught me that the best stories might be hidden inside the things we don't usually bother to see—from smoke alarms to zippers. Most important, his assignments put me on the spelunking expedition that led to this book.

I also owe a great debt to *New York Times* editor Sheila Glazer and fact checker Steven Stern, who collaborated on the "Who Made That?" column, caught my mistakes, and built a safety net underneath me.

Eamon Dolan, my editor at Houghton Mifflin Harcourt, read draft after draft of the manuscript; he became my GPS system as I mined the fields of economics, psychology, engineering, ethnography, and human-factors design to understand how invention works. With great tact and charm, Eamon convinced me to climb out of the rabbit holes that I fell into while trying to answer certain baffling questions. The imprint of his mind is on every page of this book.

My agent, David McCormick, was wonderfully supportive of this project in its infancy; he helped me turn a hunch into a thesis and then a thirty-page proposal.

I'd also like to thank Team Inventology—the freelance researchers, transcribers, translators, and editors who helped prepare this book for publication. These hired guns aided me as I verified quotes, tracked down obscure studies, puzzled over Russian manuscripts, and combed through every sentence in an attempt to catch any errors that might have crept into the text. My most stalwart fact checkers were Whitney Light, Tracy Walsh, Jesse Marx, Allan Guzman, Lillian Steenblik Hwang, Aimee Kuvadia, and Jessica Johnson. Jesse Marx assisted me with photo research and turned up some amazing images. Allan Guzman and James Pouliot transcribed many of the interviews that appear in this book. Thomas Kitson provided soulful Russian-to-English translations. Lisa Dierbeck, Jen Block, and Karen Propp proffered illuminating feedback and editing advice.

Finally, I'd like to acknowledge the many inventors who generously agreed to open up their lives and minds to me. This book is a love letter to them.

NOTES

Introduction

Page

ix *created aviator sunglasses:* Pagan Kennedy, "Who Made Those Aviator Sunglasses?," *New York Times Magazine,* August 3, 2012, http://www.nytimes.com/2012/08/05/magazine/who-made-those-aviator-sunglasses.html?_r=0.

led to the Xerox machine: "The Story of Xerography," Xerox pamphlet, August 9, 1999, http://www.xerox.com/downloads/usa/en/s/Storyofxerography.pdf.

retrieving hundreds of balls: The details and quotes in this section are drawn from an interview with Jake Stap by the author, July 2012.

x *"'I could have thought of that'":* Sue Kust, interview with the author, July 2012.

standard tennis equipment: Jan R. Magnus and Franc J. G. M. Klaassen, "The Effect of New Balls in Tennis: Four Years at Wimbledon," *The Statistician* 48 (1999): 239–46, http://www.researchgate.net/profile/Franc_Klaassen/publication/228582046_Thc_effect_of_new_balls_in_tennis_four_years_at_Wimbledon/links/0c96051a7631910d08000000.pdf.

xi *"the creation becomes an invention":* Art Fry, "Creativity, Innovation and Invention: A Corporate Inventor's Perspective," *CPSB's Communiqué* 13 (2002): 1–5, http://www.cpsb.com/research/communique/new-product-development/Art-Frey-Creativity-and-Innovation.pdf.

"operate the device in my mind": Nikola Tesla, *My Inventions: The Autobiography of Nikola Tesla* (Soho Books, no date or location given), 13.

xii *Eindhoven:* William Pentland, "World's 15 Most Inventive Cities," *Forbes* website, posted July 9, 2013, http://www.forbes.com/sites/williampentland/2013/07/09/worlds-15-most-inventive-cities/.

xv *Kekulé's famous dream:* Malcolm W. Browne, "The Benzene Ring: Dream Analysis," *New York Times,* August 16, 1988, http://www.nytimes.com/1988/08/16/science/the-benzene-ring-dream-analysis.html.

a theatrical stunt: Martin Cooper, interviews and e-mail exchanges with the author, 2013–14.

xvi *Alexander Fleming returned:* Dr. Howard Markel, "The Real Story Behind Penicillin," *PBS NewsHour, The Rundown* (blog), September 27, 2013, http://www.pbs.org/newshour/rundown/the-real-story-behind-the-worlds-first-antibiotic/. See also Alexander Fleming, "On the Antibacterial Action of Cultures of a Penicillium, with Special Reference to Their Use in the Isolation of *B. influenzae*," *British Journal of Experimental Pathology* 10 (1929): 226–36.

1. Martian Jet Lag

3 *noticed a workman pushing:* Joe Sharkey, "Reinventing the Suitcase by Adding the Wheel," *New York Times*, October 4, 2010, http://www.nytimes.com/2010/10/05/business/05road.html?_r=0.

pulled it behind you: Bernard Sadow, U.S. Patent 3653474—"Rolling Luggage," application filed February 16, 1970, and issued on April 4, 1972.

sold well in the 1970s: Corey Kilgannon, "From Suitcases on Wheels to Tear-Free Onion Slicers," *New York Times*, August 6, 2000, http://www.nytimes.com/2000/08/06/nyregion/from-suitcases-on-wheels-to-tear-free-onion-slicers.html.

Sadow's design: I deduced this from the blueprint in Sadow's patent.

rigid handle: Robert V. Plath, U.S. Patent 4995487—"Wheeled Suitcase and Luggage Support," application filed August 8, 1989, and issued February 26, 1991.

4 *without attacking your ankles:* Bruce Horovitz, "Success Stunned Pilot Inventor," *USA Today*, February 19, 2003, http://usatoday30.usatoday.com/money/biztravel/2003-02-19-bagside_x.htm.

"Life Sucks and Then You Fly": Warren Berger, "Life Sucks and Then You Fly," *Wired*, August 1, 1999, 156–63.

Rollaboard suitcase took off: Scott Applebee, "Travelpro Announces Its Celebration of the 25th Anniversary of the Invention of the Rollaboard," *Travelpro* (blog), January 10, 2012, http://travelproluggageblog.com/tag/robert-plath/.

tasks that we repeat: Adam Smith, *An Inquiry into the Nature and Causes of the Wealth of Nations*, Project Gutenberg reprint, uploaded in 2009, http://www.gutenberg.org/files/3300/3300-h/3300-h.htm.

5 *boring job:* Eric von Hippel, "Democratizing Innovation," MIT World, video of an April 2005 lecture, http://video.mit.edu/watch/democratizing-innovation-9940/.

emoticon: Pagan Kennedy, "Who Made That Emoticon?" *New York Times Magazine*, November 23, 2012, http://www.nytimes.com/2012/11/25/magazine/who-made-that-emoticon.html.

6 *"Can we . . . produce it for them?":* Eric von Hippel, interview with Tom Austin for Gartner, Inc., "Eric von Hippel on Democratizing Innovation," *Blogging on Business* (blog), September 23, 2013, http://bobmorris.biz/eric-von-hippel-on-democratizing-innovation-an-interview-by-tom-austin-for-gartner-inc.

80 percent of the scientific equipment products: Eric von Hippel, *The Sources of Innovation*, originally published in 1988, Creative Commons License for open distribution online, 11–18, http://web.mit.edu/evhippel/www/sources.htm.

holds true in many other fields: Eric von Hippel, "Development of Products by Lead Users," in *Democratizing Innovation*, originally published in 2005, now distributed under a Creative Commons license online, http://web.mit.edu/people/evhippel/books/DI/Chapter2.pdf. Von Hippel and his colleagues found that the origin of inventions varied wildly from industry to industry. For instance, when they studied improvements in tractor-shovel design, they found that employees at manufacturing companies had come up with most of the key breakthroughs; so in that industry, Lead Users contributed very little. However, when it came to pultrusion tools (equipment that makes customized objects out of polymers), the Lead Users were the ones who contributed nearly all the important breakthroughs. Of course, it's important to note that much of this research was conducted decades ago, at a time when the barriers to entry were much higher for user-inventors than they are now.

7 *"an integral part of their culture":* Nat Sims, interview with the author, 2013.

"need-forecasting laboratory": Eric von Hippel, "Lead Users: A Source of Novel Product Concepts," *Management Science* 32, no. 7 (July 1986): 791–801.

Martian jet lag: William Gibson, *Distrust That Particular Flavor* (New York: G. P. Putnam's Sons, 2012), 99.

scientist, Deborah Bass: Rebecca Boyle, "How Do You Tell Time on Mars?" *Popular Science*, March 9, 2012, http://www.popsci.com/science/article/2012-03/how-do-you-tell-time-mars.

MarsClock phone app: Scott Maxwell, e-mail correspondence with the author, 2015.

8 *developed their own lingo:* Jack Dorsey, "The Power of Curiosity and Inspiration," a talk given at Stanford University, February 9, 2011, http://ecorner.stanford.edu/authorMaterialInfo.html?mid=2635.

9 *"I was considering what I wanted":* D. T. Max, "Two-Hit Wonder," *The New Yorker*, October 21, 2013, http://www.newyorker.com/magazine/2013/10/21/two-hit-wonder.

(watching the bison): Dorsey's Stanford talk, February 9, 2011.

10 *site was awkward to use:* Nick Bilton, *Hatching Twitter: A True Story of Money, Power, Friendship, and Betrayal* (New York: Penguin Group, 2013), Kindle edition, 117.

@ sign: Garrett Murray, "The Real History of the @reply on Twitter," *Maniacal Rage* (blog), July 10, 2012, http://log.maniacalrage.net/post/26935842947/the-real-history-of-the-reply-on-twitter.

"all the early tweets": Max, "Two-Hit Wonder."

11 *hashtag and the retweet:* Zachary M. Seward, "The First-Ever Hashtag, @-Reply and Retweet, as Twitter Users Invented Them," *Quartz* (online magazine), Oc-

tober 15, 2013, http://qz.com/135149/the-first-ever-hashtag-reply-and-retweet-as-twitter-users-invented-them/.

2. User-Inventors

12 *"It weighed ninety pounds":* Tim Derk, interview with the author, 2013.

13 *"You've got to look out of the eyes":* Kenn Solomon, interview with the author, 2013.

"negative spaces": Elizabeth L. Rosenblatt, "A Theory of IP's Negative Space," *Columbia Journal of Law & the Arts* 34, no. 3 (2011): 317.

emerged from the negative spaces: Ibid., 442–43.

14 *"dark matter":* Quoted in Patricia Cohen, "Innovation Far Removed from the Lab," *New York Times*, February 9, 2011, http://www.nytimes.com/2011/02/10/arts/10innovative.html.

dwarfs the efforts: Eric von Hippel et al., "Comparing Business and Household Sector Innovation in Consumer Products: Findings from a Representative Study in the UK," *Social Science Research Network*, September 27, 2010, http://papers.ssrn.com/sol3/papers.cfm?abstract_id=1683503.

track spending on R&D: "National Patterns of R&D Resources: 2011–12 Data Update," National Science Foundation, December 2013, http://www.nsf.gov/statistics/nsf14304/content.cfm?pub_id=4326&id=1.

15 *survey of British citizens:* Stephen Flowers et al., *Measuring User Innovation in the UK*, NESTA Research Report, April 2010, http://www.nesta.org.uk/sites/default/files/measuring_user_innovation_in_the_uk.pdf.

2013 Time *magazine survey:* "The TIME Invention Poll," produced in cooperation with Qualcomm, November 14, 2013, https://www.qualcomm.com/documents/time-invention-poll-survey-data-full-report.

doctor friends: Nat Sims, interview with the author, 2013.

16 *"cable that ran down my arm":* Debra Latour, interviews and correspondence with the author, 2014–15.

17 *people with upper-body differences:* Debra Latour, "The Ipsilateral Scapular Cutaneous Anchor: Implications in Consumer Use," American Academy of Orthotists & Prosthetists, 37th Academy Annual Meeting and Scientific Symposium, March 2011, http://www.oandp.org/publications/jop/2011/2011-59.pdf.

passion to help other people: Latour, interview with the author, 2014.

Superhero Cyborgs workshop: Eight of the ten kids who enrolled in the Superhero Cyborgs workshop were able-bodied; they'd signed up for the program because of their interest in engineering and design. This had surprised the organizers of the camp, who expected to cater to children with upper-body limitations. As a result, many of the campers contemplated the fascinating problem of creating prosthetics for both the able-bodied and those with body differences—for instance, is a phone a prosthetic device? And if so, what is the best way to attach it to the body?

18 *"simplest ideas are the very best":* These details are drawn from my reporting at the camp during the summer of 2014.

$7,000 for a hand: Latour pointed out that the Shriners Hospital for Children provides all prosthetic devices at no cost to the family; if the insurance company won't pay, the hospital waives the fee. However, many families do not live near a hospital that provides this service, and so they may not have access to low-cost or free prostheses.

conference where engineering students: "A New Hand for Lucas," Rochester Institute of Technology website, no date given, http://www.rit.edu/showcase/index.php?id=252.

19 *"Are we doing anything to help them?":* Jon Schull, interviews and e-mails with the author, 2014.

under $30: Steve Henn and Cindy Carpien, "3-D Printer Brings Dexterity to Children with No Fingers," *National Public Radio,* June 18, 2013, http://www.npr.org/blogs/health/2013/06/18/191279201/3-d-printer-brings-dexterity-to-children-with-no-fingers.

20 *seven hundred prosthetic devices:* e-NABLE community on Google+, https://plus.google.com/communities/102497715636887179986.

"And why not?": Schull, interview with the author.

21 "They *should make a milk carton":* Tip of the hat to Eric von Hippel for addressing this idea in his 1988 book. He talks a great deal about our misconceptions about creativity and invention.

prosumer*:* "Prosumer," *Oxford Dictionaries,* definition accessed November 12, 2014, http://www.oxforddictionaries.com/us/definition/american_english/prosumer.

3. Someone Else's Shoes

23 *less energy than the stairs:* Amy Smith's quotes and related details come from my reporting in 2003. Some of the material in this chapter appeared in the *New York Times Magazine* on November, 30, 2003, in an article titled "Necessity Is the Mother of Invention." In addition, I corresponded with Smith in 2015 in order to be able to add updates and new details to this story.

24 *"trying to solve hunger":* Mark Belinsky, interviews and e-mails with the author, 2013–14.

"You can do a lot more testing": Amy Smith, from a tape recording of her class that I made in 2003.

25 *PortaTherm:* M. Dominguez et al., "The MIT D-Lab Electricity-Free PortaTherm™ Incubator for Remote Testing with the QuantiFERON®-TB Gold In-Tube Assay," *International Journal of Tuberculosis and Lung Disease,* November 2010, 1468.

26 *"It got sickening":* Jamy Drouillard, interview with the author, 2003.

Pythagorean theorem at the dinner table: Amy Smith, interview with the author, 2003.

27 *"hair up in bandannas":* Ibid.

28 *pot-within-a-pot:* Rolex Awards for Enterprise, "Mohammed Bah Abba, 2000 Laureate, Applied Technology," http://www.rolexawards.com/profiles/laureates/mohammed_bah_abba.

29 *"technology can affect the well-being of a people":* Shawn Frayne, interview with the author, 2003.

"saw a flag": All of the post-2003 details and quotes related to Frayne come from my interviews and e-mail correspondence with him in 2014 and 2015.

30 *"Why are you doing that?":* Abbie Griffin, Raymond Price, and Bruce Vojak, *Serial Innovators: How Individuals Create and Deliver Breakthrough Innovations in Mature Firms* (Stanford, CA: Stanford University Press, 2012), Kindle edition, locations 1305–16.

"when someone cares": Frayne, interview with the author, 2014.

"inside the mind of the user": Martin Cooper, interview with the author, 2014.

4. The Future of Feedback

31 *"Dick's Book of Dumb Ideas":* Dick Belanger, phone interviews and follow-up with the author, 2013–14. He also wrote a short "memoir" for the author to help flesh out the details of this story.

34 *"continuous improvement":* Gary Klein, interview with the author, 2014.

35 *"potential problems":* Gary Klein, "Performing a Project Premortem," *Harvard Business Review,* September 2007, https://hbr.org/2007/09/performing-a-project-premortem.

37 *"more important than 3-D printing":* Frayne, interview with the author, 2014.

"raised $150,000": The Kickstarter page for the product can be found at https://www.kickstarter.com/projects/jmcrae/b-squares-modular-solar-powered-electrics. Actually, Frayne and his collaborators raised $145,034, but I left the $150,000 in the quote because it seemed fair to round up the number.

nine months: Mark Belinksy, interviews with the author, 2013–14.

39 *more than $70,000:* I gathered this information first from Belinsky and then confirmed it by checking the Indiegogo page.

"think about tornadoes": Belinsky, interview with the author, 2014.

40 *"I don't want to be cheesy":* Michael D. Greenberg and Elizabeth M. Gerber, "Learning to Fail: Experiencing Public Failure Online Through Crowdfunding," in *Proceedings of the SIGCHI Conference on Human Factors in Computing Systems* (New York: ACM, 2014), 581–90.

award-winning designer: Gregory Ng, "Is the PowerSquid Surge Protector Worth It?" *Apple Matters,* May 2, 2006, http://www.applematters.com/article/is-the-powersquid-surge-protector-worth-it/.

"Was it in the cost?": Chris Hawker, interview with the author, 2014.

41 *quote attributed to Henry Ford:* It's unclear whether Ford actually uttered this statement; still, the quip nicely captures his philosophy. See "My Customers

Would Have Asked for a Faster Horse," *Quote Investigator* (blog), July 28, 2011, http://quoteinvestigator.com/2011/07/28/ford-faster-horse/.

42 *"five years of waiting":* Hawker, interview with the author, 2014.

5. Super-Encounterers

47 *"But the nozzle that was different":* Lonnie Johnson, interview with the author, 2013.

Johnson spent years: Ibid. See also Pagan Kennedy, "Who Made That Super Soaker?" *New York Times Magazine*, August 2, 2013, http://www.nytimes.com/2013/08/04/magazine/who-made-that-super-soaker.html?_r=0.

48 *"sugar on my hands":* Quoted in Robert H. Mazur, "Discovery of Aspartame," in *Aspartame: Physiology and Biochemistry* (Boca Raton, FL: CRC Press, 1984), 4.

49 *sucralose:* Walter Gratzer, *Eurekas and Euphorias: The Oxford Book of Scientific Anecdotes* (Oxford: Oxford University Press, 2002), Google e-book version, 37–38.

"frontier of knowledge": Sunny Y. Auyang, "Chance and the Prepared Mind in Drug Discovery," *Creating Technology* (website that posts research on the history of science), http://www.creatingtechnology.org/biomed/chance.htm.

European study of thousands of inventors: Alfonso Gambardella et al., "The Value of European Patents: Evidence from a Survey of European Inventors," *Final Report of the PatVal EU Project*, January 2005, http://ec.europa.eu/invest-in-research/pdf/download_en/patval_mainreportandannexes.pdf.

cigarette in Pearsall's workroom: David A. Lucht, "Where There's Smoke," *NFPA Journal*, March/April 2013, 51–52, http://bit.ly/1uQQq33.

Honeywell: The company was then called the Minneapolis-Honeywell Regulator Company.

50 *"Cut the static crap":* Lucht, "Where There's Smoke," 55–56.

"But it did": Duane Pearsall, from his unpublished memoir, quoted in ibid., 55.

heat-sensing alarms: John E. Lindberg, U.S. Patent 3429183—"Temperature Averaging System," filed on July 14, 1966. Also see John E. Lindberg, U.S. Patent 3896423—"Fire and Overheat Detection System," filed on September 14, 1973.

51 *Graham Wallas:* Graham Wallas, *The Art of Thought* (Kent, England: Solis Press, 2014), reprint of the 1926 book.

52 *"'Oh, my God, this is amazing'":* Steve Hollinger, interviews, reportage, and correspondence with the author, 2013–14.

53 *"the knack of finding things":* Auyang, "Chance and the Prepared Mind in Drug Discovery."

54 *count the photographs:* Richard Wiseman, "Be Lucky—It's an Easy Skill to Learn," *Telegraph*, January 9, 2003, http://www.telegraph.co.uk/technology/3304496/Be-lucky-its-an-easy-skill-to-learn.html.

55 *Super-Encounterers:* Sanda Erdelez, "Information Encountering: It's More

Than Just Bumping into Information," *Bulletin of the American Society for Information Services*, February/March 1999, http://www.asis.org/Bulletin/Feb-99/erdelez.html.

"leave your house": Stephann Makri and Ann Blandford, "Coming Across Information Serendipitously—Part 1: A Process Model," *Journal of Documentation* 68, no. 5 (2012): 684–705, http://www.emeraldinsight.com/doiabs/10.1108/00220411211256030.

the music of the street: Stephann Makri et al., "'Making My Own Luck': Serendipity Strategies and How to Support Them in Digital Information Environments," *Journal of the Association for Information Science and Technology* 65, no. 11 (2014): 2179–94, http://onlinelibrary.wiley.com/doi/10.1002/asi.23200/abstract.

56 *"a chat with anyone":* Stephann Makri, interview with the author, 2014.

57 *buffet of objects:* Jacob W. Getzels and Mihaly Csikszentmihalyi, *The Creative Vision: A Longitudinal Study of Problem Finding in Art* (New York: John Wiley, 1976), 85.

before he drew them: All of the students in the study were male.

"I felt it alive": Getzels and Csikszentmihalyi, *The Creative Vision*, 153.

58 *this* one-hour *test:* Though no formal time restriction was placed on the students, they were encouraged to finish their drawings in about an hour.

the word tinker*:* "Tinker," *Oxford Dictionaries*, definition accessed December 21, 2014, http://www.oxforddictionaries.com/us/definition/american_english/tinker.

59 *it was stinky:* J. Kim Vandiver and Pagan Kennedy, "Harold E. Edgerton (1903–1990)," *Biographical Memoirs* 86 (Washington, DC: National Academies Press, 2005), http://edgerton-digital-collections.org/wp-content/uploads//hedgerton-nas-memoir.pdf. Note: Some of the text pertaining to Edgerton has been adapted from that article—that is, from earlier writings by the author.

61 *"brain in a dish":* Nancy J. Nersessian, "How Do Engineering Scientists Think? Model-Based Simulation in Biomedical Engineering Research Laboratories," *Topics in Cognitive Science* 1, no. 4 (2009): 730–57, http://onlinelibrary.wiley.com/doi/10.1111/j.1756-8765.2009.01032.x/full.

"Things go wrong all the time": Nancy Nersessian, interview with the author, 2013.

62 *"a simple truth":* Charles H. Townes, *How the Laser Happened: Adventures of a Scientist* (Oxford: Oxford University Press, 1999), retrieved from Google Books.

6. Data Goggles

63 *"None of us at Pfizer":* Ian Osterloh, "How I Discovered Viagra," *Cosmos*, July 18, 2007, http://cosmosmagazine.com/features/how-i-discovered-viagra/.

"targeted searches": Adam Matthias, "Integrating Research and Development: The Emergence of Rational Drug Design in the Pharmaceutical Industry,"

Studies in History and Philosophy of Biological and Biomedical Sciences 36, no. 3 (September 2005): 513–37, http://www.ncbi.nlm.nih.gov/pubmed/16137601.

64 *"decline in serendipity":* Alan A. Baumeister et al., "On the Exploitation of Serendipity in Drug Discovery," *Clinical & Experimental Pharmacology* 3, no. 121, published online under a Creative Commons license at http://omicsonline.org/on-the-exploitation-of-serendipity-in-drug-discovery-2161-1459.1000e121.php?aid=14387.

65 *$100 billion* every year: Jamie Cattell et al., "How Big Data Can Revolutionize Pharmaceutical R&D," McKinsey & Company, April 2013, http://www.mckinsey.com/insights/health_systems_and_services/how_big_data_can_revolutionize_pharmaceutical_r_and_d.

"The time will soon come": Steve Dickman, "Big Data in Drug Discovery and Healthcare: What Is the Tipping Point?" *Boston Biotech Watch,* January 23, 2014, http://bostonbiotechwatch.com/2014/01/23/big-data-in-drug-discovery-and-healthcare-what-is-the-tipping-point/.

66 *"Oh, my God, a naked lady!":* Murray Robinson. The quotes and details related to the Robinsons' childhood, as well as details about Murray Robinson's work life, are drawn from interviews and correspondence with the author in 2013–15.

67 *Ann Smith examined an infant:* John I. Gallin, *Director's Annual Report,* NIH Clinical Center, 2009, http://clinicalcenter.nih.gov/about/profile/pdf/Profile_2009.pdf.

68 *one in every twenty-five thousand people:* Ann Smith, e-mail correspondence with the author, 2015.

lucky break: Ibid.; and Mary Jane Fine, "Making Sense of Disorder Delco Center Finds Retardation Clues," *Philadelphia Inquirer,* November 19, 1991, http://articles.philly.com/1991-11-19/news/25769061_1_mental-retardation-genetic-disorders-neon-signs.

69 *"next big problem in my field":* Robinson, interviews and correspondence with the author, 2013–15.

70 *gene chip:* Atul Butte, "Translating a Trillion Points of Data," keynote presentation at the 9th International Digital Curation Conference, San Francisco, CA, February 25, 2014.

71 *measured in petabytes:* Vivien Marx, "Biology: The Big Challenges of Big Data," *Nature,* June 12, 2013, http://www.nature.com/nature/journal/v498/n7453/full/498255a.html.

in the Library of Congress: Brian McKenna, "What Does a Petabyte Look Like?" *Computer Weekly,* no date, http://www.computerweekly.com/feature/What-does-a-petabyte-look-like.

"new uses of old drugs": Jiao Li et al., "A Survey of Current Trends in Computational Drug Repositioning," *Briefings in Bioinformatics Advance Access,* March 31, 2015, http://bib.oxfordjournals.org/content/early/2015/03/31/bib.bbv020.full.pdf.

72 *shrink cancer tumors:* Atul Butte, interview and correspondence with the author, 2014–15.

"our work shows the power": Nadine S. Jahchan et al., "A Drug Repositioning Approach Identifies Tricyclic Antidepressants as Inhibitors of Small Cell Lung Cancer and Other Neuroendocrine Tumors," *Cancer Discovery,* December 2013, 1364, http://cancerdiscovery.aacrjournals.org/content/3/12/1364.long.

data to clinical trials in about two years: "FDA-Approved Antidepressant May Combat Deadly Form of Lung Cancer, Study Finds," Stanford Medicine News Center, September 26, 2013, http://med.stanford.edu/news/all-news/2013/09/fda-approved-antidepressant-may-combat-deadly-form-of-lung-cancer-study-finds.html.

preeclampsia: Adam Bonislawski, "Carmenta Bioscience to Take Preeclampsia Test, Streamlined Dx Development Approach to Progenity," *GenomeWeb,* April 29, 2015, https://www.genomeweb.com/proteomics-protein-research/carmenta-bioscience-take-preeclampsia-test-streamlined-dx-development.

"it's not like the air": Butte, interview and correspondence with the author, 2014–15.

74 *Assay Depot:* Brian Fung, "Have You Heard of Assay Depot? It's the Amazon.com of Medical Research," *Atlantic,* April 13, 2012, http://www.theatlantic.com/health/archive/2012/04/have-you-heard-of-assay-depot-its-the-amazon com-of-medical-research/255871/.

75 *"right off the Internet":* Butte, interview with the author, 2014.

"your source of cells": Advanced Tissue Services, http://advancedtissueservices.com/.

"why not in the future?": Butte, interview with the author, 2014.

begin investing in pharmaceutical companies: "Request for Startups," Y Combinator, September 2014, https://www.ycombinator.com/rfs/.

"two postdocs with a laptop": Max Hodak, "Adding Potential Energy: Transcriptic's Series A," *Transcriptic* (blog), February 19, 2015, https://www.transcriptic.com/blog/2015/02/19/series-a.html.

76 *two drugs, taken together:* Yogen Saunthararajah, interview and extended correspondence with the author, 2014–15.

patients saw improvements: Yogen Saunthararajah et al., "Evaluation of Noncytotoxic DNMT1-Depleting Therapy in Patients with Myelodysplastic Syndromes," *Journal of Clinical Investigation* 125, no. 3 (2015): 1058, http://www.jci.org/articles/view/78789/pdf.

77 *"It takes artistry":* Saunthararajah, correspondence with the author, 2015.

"frozen knowledge": Butte, interview with the author, 2014.

7. Building an Empire Out of Nothing

79 *waited for it to die:* Much of this is based on material in Michael Kantor and Laurence Maslon, *Make 'Em Laugh: The Funny Business of America* (New York: Grand Central Publishing, 2008), accessed on Google Books.

"show about nothing": Seinfeld, Season 4, "The Pitch to NBC," YouTube, uploaded December 6, 2011, https://www.youtube.com/watch?v=ofOSlsNz5I8.

80 *the same color book:* Lawrence Herbert, interviews with the author, 2013.

81 *a million dollars a year:* Pagan Kennedy, "Who Made That Pantone Chip?" *New York Times Magazine,* February 22, 2013, http://www.nytimes.com/2013/02/24/magazine/who-made-that-pantone-chip.html.

consumer tastes: Herbert, interview with the author, 2014.

millions of dollars a year: Jude Stewart, "Pantone Expands from Color Services to Hotels (with Rooms to Dye For)," *Fast Company,* May 19, 2010, http://www.fastcompany.com/1649699/pantone-expands-color-services-hotels-rooms-dye.

82 *"excess of meaning":* Alice Flaherty, interviews and correspondence with the author over many years. I also drew details from her book, *The Midnight Disease: The Drive to Writer, Writer's Block, and the Creative Brain* (Boston: Houghton Mifflin, 2004).

83 *"The itchy guy":* From the author's interviews with Flaherty.

84 *"novel but useless":* Alice W. Flaherty, "Brain Illness and Creativity: Mechanisms and Treatment Risks," *Canadian Journal of Psychiatry* 56, no. 3 (March 2011): 132–43, https://www.questia.com/library/journal/1P3-2349531741/brain-illness-and-creativity-mechanisms-and-treatment.

"had need for copies": Chester Carlson, quoted in David T. Kearns and David A. Nadler, *Prophets in the Dark: How Xerox Reinvented Itself and Beat Back the Japanese* (New York: HarperCollins, 1992), 5.

"the invention was too promising to be dormant": "The Story of Xerography."

85 *Ted Baker and Reed E. Nelson:* Ted Baker and Reed E. Nelson, "Creating Something from Nothing: Resource Construction Through Entrepreneurial Bricolage," *Administrative Science Quarterly* 50, no. 3 (2005): 329–66, http://asq.sagepub.com/content/50/3/329.

87 *harvest its garbage:* Tim Anderson, phone conversations and correspondence with the author, 2014–15.

88 *one of the most prolific hubs:* Philip J. Hilts, "Last Rites for a 'Plywood Palace' That Was a Rock of Science," *New York Times,* March 31, 1998.

fewer than five hundred 3-D printers: Terry Wohlers, "The World of Rapid Prototyping," *Proceedings of the Fourth International Conference on Desktop Manufacturing,* September 1992, http://wohlersassociates.com/mr.html.

89 *"inside of a solid sphere":* Tim Anderson, phone conversations with the author, 2014.

Shrunken Head Machine: Jim Bredt, interview and correspondence with the author, 2014–15.

90 *"looking like a big turd":* Tim Anderson and Jim Bredt, interviews with the author, 2014.

91 *fastest 3-D printer:* Marina Hatsopoulos, "Z Corp. to Launch World's Fastest 3D Printer at Autofact 97," Z Corporation, October 30, 1997, http://www.rp-ml.org/rp-ml-1997/2526.html.

a leader in the nascent industry: Terry Wohlers, "1997: A Year of News, Surprises and Challenges," *Prototyping Technology International*, 1998, http://www.wohlersassociates.com/98pti.html.
most-patented inventor: Michael Molitch-Hou, "Report Compiles, Categorizes and Makes Pretty 3D Printing Patent Data," *3D Printing Industry*, March 20, 2014, http://3dprintingindustry.com/2014/03/20/3d-printing-patent-data-report/.

92 *"figured out how to make synthetic ammonia":* Bill Gates, "Here's My Plan to Improve Our World—and How You Can Help," *Wired* magazine, posted online November 12, 2013, http://www.wired.com/2013/11/bill-gates-wired-essay/.
lives saved: Fritz Haber and Carl Bosch's legacy as humanitarians is tainted by their record as inventors of weapons of war. During World War I, Haber headed up the German chemical-research effort and contributed to the development of weapons—like chlorine gas—that had been outlawed by the Hague Convention of 1907. Bosch also contributed to the design and manufacture of German munitions in the early twentieth century.
"liable to have the greatest impact": Gates, "Here's My Plan to Improve Our World."

8. The Pong Effect

97 *"I have a mantra":* Martin Cooper, interviews and e-mail correspondence with the author, 2013–14.

98 *began in 1955:* "Handie-Talkie Paging System," a timeline on the Motorola website that shows the evolution of the company's hand-held communicators, http://www.motorolasolutions.com/en_xl/about/company-overview/history/explore-motorola-heritage/handie-talkie-paging-system.html.

99 *listen to her child in another room:* Details about the hospital loudspeakers and the Motorola system come from *The Handie-Talkie Pager Newsreel*, an industrial movie made by Motorola in 1965, YouTube, uploaded June 28, 2011, https://www.youtube.com/watch?v=rHqLixYhbXs.
suitcase-size box: "The History of Ericsson," company timeline, http://www.ericssonhistory.com/timeline/1956/.
Atomic Light Unit: "Dick Tracy Stories—1969," *Dick Tracy Depot* (blog), http://dicktracy.info/comic-strip/storylines/dick-tracy-stories-1969/.

100 *hurled his against the wall:* Cooper, interview with the author.
then a product manager at Motorola: Martin Cooper, interview with Sean Maloney, oral history reference number X4602.2008, Computer History Museum, Mountain View, CA, May 2, 2008, 9.
"could not conduct the business": Cooper, interview with the author.

101 *"like a backpack":* Ibid.
mobile phone booths: When I talked to Cooper, he was working on his autobiography and had just spent months going through every Bell Labs document

related to mobile phones to see whether AT&T had put any effort into developing a hand-held cell phone. He told me that there had been no discussion (on paper) of a hand-held phone. Of course, it's hard to verify that without a complete scouring of the archive. AT&T's strategy in the 1960s does suggest that the behemoth had bet on car phones and did not believe that people would wear their phones on their bodies.

102 *beaming signals to individual cars:* Edward Joel Amos Jr., U.S. Patent 3663762—"Mobile Communication System," application filed on December, 21, 1970, and issued on May 16, 1972. Also, an AT&T industrial film from 1978 describes field experiments in which its cellular towers beam out signals to cars and trucks—and not to hand-held phones. See https://www.youtube.com/watch?v=d6X_1PcR_gs.

encourage the upstarts: "Motorola Demonstrates Portable Telephone," Motorola Inc. Communications Division press release, April 3, 1973, http://demandware.edgesuite.net/aahb_prd/on/demandware.static/Sites-Motorola_US-Site/Sites-Motorola_US-Library/en_US/v1364939238252/corporate/about-us/docs/Motorola-Heritage-DynaTAC-NewsRelease.pdf.

"The worst possible thing": Cooper, Computer History Museum interview, 11.

"like science fiction": Cooper, interview with the author.

103 *"literally shut down":* Cooper, Computer History Museum interview, 11.

Bob Galvin: Cooper, interview with the author.

"What's that?": Cooper, Computer History Museum interview, 11.

from an episode of The Jetsons*:* Cooper sent me photos of all the phone designs that the team produced.

"called the 'brick'": Cooper, interview with the author.

104 *Dick Tracy gadgets:* This description comes from the author's search through eBay for images of vintage Dick Tracy toys.

105 *"how things could be done differently":* Cooper, interview with the author.

106 *"We shape our tools":* J. M. Culkin, "A Schoolman's Guide to Marshall McLuhan," *Saturday Review*, March 18, 1967, 51–53, 71–72.

"primitive prototype is typical": Tim Wu, *The Master Switch: The Rise and Fall of Information Empires* (New York: Knopf Doubleday Publishing Group, 2010), Kindle edition, locations 2501–8.

9. The Wayne Gretzky Game

108 *"future vice-president":* Vannevar Bush, "The Inscrutable Thirties," in *From Memex to Hypertext*, ed. James M. Nyce and Paul Kayn (Boston: Academic Press, 1991), 67.

109 *stored in a desk drawer:* It's not clear when Bush first conceived of the Memex, though in his autobiographical writings he puts the date at 1932.

"scramble for money": Colin Burke, "A Practical View of Memex: The Career of the Rapid Selector," in *From Memex to Hyptertext*, ed. Nyce and Kayn, 145.

110 *"my Aunt Susie":* Vannevar Bush, "Of Inventions and Inventors," an excerpt from his autobiography republished in *From Memex to Hypertext*, ed. Nyce and Kayn, 226.

111 *"take a long look ahead":* Vannevar Bush, "Memex II," an essay republished in *From Memex to Hypertext*, ed. Nyce and Kayn, 166.

one morning in 1950: Some sources cite 1951 as the date of Engelbart's engagement; others cite 1950. Engelbart's *New York Times* obituary specified the latter date, so I went with that version since it seemed to be the most reliable one. John Markoff, "Computer Visionary Who Invented the Mouse," *New York Times*, July 3, 2013, http://www.nytimes.com/2013/07/04/technology/douglas-c-engelbart-inventor-of-the-computer-mouse-dies-at-88.html.

a crusade: Douglas Engelbart, interviewed by Judy Adams and Henry Lowood in the 1980s, archived in the Stanford library system.

112 *"Your whole body was involved":* Christina Engelbart, interview with the author, 2013.

113 *"English Lit majors":* Doug Engelbart, Stanford interview.

"you're not . . . going to last": Christina Engelbart, interview with the author.

114 *sore neck:* Bill English, interview with the author, 2013. English worked at SRI during that time.

115 *"trouble understanding what he was doing":* Cade Metz, "The Mother of All Demos—150 Years Ahead of Its Time," *The Register*, December 11, 2008, http://www.theregister.co.uk/2008/12/11/engelbart_celebration/.

high fever: Cade Metz, "Doug Engelbart, Who Foresaw the Modern Computer, Dies at 88," *Wired*, July 3, 2013, http://www.wired.com/2013/07/douglas_engelbart_mouse/.

"like Moses opening the Red Sea": Tia O'Brien, "From the Archives: Douglas Engelbart's Lasting Legacy, 1999," *San Jose Mercury News*, February 9, 1999, http://www.mercurynews.com/ci_23592605/mouse-man-douglas-engelbart-early-ideas-computing.

Wayne Gretzky Game: Alan Kay, "The Future Doesn't Have to Be Incremental," a lecture given at the DEMO Enterprise conference, 2014, https://www.youtube.com/watch?v=gTAghAJcO1o.

117 *"The men had no background":* Douglas K. Smith and Robert C. Alexander, *Fumbling the Future* (New York: William Morrow, 1988), 209.

118 *"You could change the world!":* Benjamin Mayo, "PARC Scientist Retells Story of Jobs at Xerox: 'You're Sitting on a Gold Mine!' (video)," *9to5 Mac*, August 24, 2014, http://9to5mac.com/2014/08/24/parc-scientist-retells-story-of-jobs-at-xerox-your-sitting-on-a-gold-mine-video/.

119 *"A mosquito, as big as a human":* Valerie Landau, Eileen Clegg, and Doug Engelbart, *The Engelbart Hypothesis: Dialogs with Douglas Engelbart* (Berkeley, CA: NextPress, 2009), Kindle edition, locations 340–41.

Gordon Moore rushed up: Ibid., location 324.

120 *"predict the future in this area":* David C. Brock, "A Clear Voice: The Origins

of Gordon Moore's 1965 Paper," in *Understanding Moore's Law*, ed. David C. Brock (Philadelphia: Chemical Heritage Press, 2006), 35.

"not clear at the time": Gordon Moore, interview with Craig Addison for the Computer History Museum, January 25, 2008, http://archive.computerhistory.org/resources/access/text/2013/05/102658233-05-01-acc.pdf.

"an accurate prediction": "Excerpts from a Conversation with Gordon Moore: Moore's Law," video transcript, Intel Corporation, 2005, http://large.stanford.edu/courses/2012/ph250/lee1/docs/Excerpts_A_Conversation_with_Gordon_Moore.pdf.

121 *"ten commandments wrapped into one":* Jaron Lanier, "Musing on Moore's Law," *Wired*, May 7, 2013, http://www.wired.com/2013/05/lanier-on-moores-law/.

his coauthor: Kahn's coauthor was named Anthony Wiener. I have avoided putting his name into the text because it's so distracting. We can't see that name now without thinking of another Anthony Wiener, the one who destroyed his career by employing communications devices in a manner that no futurist in 1967 could have anticipated.

122 *80 percent accuracy rate:* Richard E. Albright, "What Can Past Technology Forecasts Tell Us About the Future?" *Technological Forecasting and Social Change*, January 2002, http://www.albrightstrategy.com/papers/Albright_Past_Forecasts.pdf.

cars would fly: Herman Kahn and A. J. Wiener, "Table XVIII: One Hundred Technical Innovations Very Likely in the Last Third of the Twentieth Century," in *The Year 2000: A Framework for Speculation on the Next Thirty-Three Years* (New York: Collier Macmillan, 1967), retrieved from http://www.crummy.com/writing/hosted/The%20Year%202000.html.

"could not stifle Moore's Law": Kevin Kelly, "Was Moore's Law Inevitable?" *The Technium* (blog), July 7, 2009, http://kk.org/thetechnium/2009/07/was-moores-law/.

123 *"a driver of what is going to happen":* "Computer History Museum Presents the 40th Anniversary of Moore's Law with Gordon Moore and Carver Mead," Computer History Museum, September 15, 2005, http://www.computerhistory.org/press/40th-anniversary-of-moore-law-with-gordon-moore-and-carver-mead.html.

"It took 20 years": Alvy Ray Smith, "How Pixar Used Moore's Law to Predict the Future," *Wired*, April 17, 2013, http://www.wired.com/2013/04/how-pixar-used-moores-law-to-predict-the-future/.

10. The Mind's R&D Lab

125 *world's first video-recording machine:* Charles P. Ginsburg, "The Development of the Ampex VTR," a memoir of his experiences at the company. The text was first delivered in slightly different form at the 82nd convention of the Society of Motion Picture and Television Engineers, October 5, 1957, and retrieved from http://www.vtoldboys.com/ampex.htm.

126 *"relationship with God":* Myron Stolaroff, *Report of Session of M. J. Stolaroff, 16 April 1956,* retrieved from https://www.erowid.org/experiences/exp.php?ID=99779.

Model 200A: John Leslie and Ross Snyder, "History of the Early Days of Ampex Corporation," AES Historical Committee, December 17, 2010, http://www.aes.org/aeshc/docs/company.histories/ampex/leslie_snyder_early-days-of-ampex.pdf.

"the greatest discovery that man has ever made": Myron J. Stolaroff, *Thanatos to Eros: 35 Years of Psychedelic Exploration* (Berlin: VWB, 1994), accessed from http://csp.org/chrestomathy/thanatos_to.html and http://www.hofmann.org/science/myron.html.

127 *fortune-telling drug:* William J. Broad, "For Delphic Oracle, Fumes and Visions," *New York Times,* March 19, 2002, http://www.nytimes.com/2002/03/19/science/for-delphic-oracle-fumes-and-visions.html.

128 *a beam of light:* Walter Isaacson, *Einstein: His Life and Universe* (New York: Simon & Schuster, 2007), Kindle edition, 26.

"I needed no models": Tesla, *My Inventions,* 12–13.

"the whole idea is worked out": James O'Neill, *Prodigal Genius* (San Diego, CA: The Book Tree, 2007), 257.

"'I expect to photograph thoughts'": Carol Bird, "Tremendous New Power Soon to Be Unleashed," *Kansas City Journal-Post,* September 10, 1933. I accessed an image of the newspaper article, as it appeared in the *Deseret News,* here: http://placefacecyberspace.net/2015/03/05/thought-camera/.

129 *"something from your mind":* Nancy Atkinson, "Elon Musk Creates Rocket Parts with the Wave of a Hand," *Universe Today* (blog), September 5, 2013, http://www.universetoday.com/104567/elon-musk-creates-rocket-parts-with-the-wave-of-a-hand/. The page includes a video of Elon Musk's demo.

130 *They were not impressed:* Stolaroff, *Thanatos to Eros,* Chapter 1, retrieved from http://www.cjayarts.com/pages/library/ThanatosToEros/c1.htm. In this section, I also relied on the account of these events by John Markoff in *What the Dormouse Said: How the Sixties Counter-Culture Shaped the Personal Computer Industry* (New York: Penguin Group, 2005).

131 *"resigned from Ampex":* Stolaroff, *Thanatos to Eros,* Chapter 1.

government regulation of psychedelics: By 1965, the FDA was charged with oversight of experiments involving psychedelic drugs, and it shut down most of the research projects in this area—including Stolaroff's. For more about the evolving legal status of psychedelic research, see Rick Doblin, *Regulation of the Medical Use of Psychedelics and Marijuana,* a dissertation submitted to the Kennedy School of Government, Harvard University, June 2001, http://www.maps.org/maps-media/332-maps-resources/research-papers/5402-dissertation-rick-doblin,-ph-d.

"There was little curiosity in our work": Stolaroff, *Thanatos to Eros.*

Stewart Brand: Fred Turner, "Stewart Brand Meets the Cybernetic Counterculture," Edge.org, October 2, 2006, http://edge.org/conversation/stewart-brand-meets-the-cybernetic-counterculture.

"Mr. Transistor": Irwin Wunderman obituary, *San Jose Mercury News,* July 26, 2005, retrieved from http://www.findagrave.com/cgi-bin/fg.cgi?page=gr&GRid=51710105.

132 *"day after day and look at the future":* Willis Harman, interview with Jeffrey Mishlove on the *Thinking Allowed* radio show, 1998, http://www.williamjames.com/transcripts/harman2s.htm.

"We brought in senior scientists": Jim Fadiman, interview with Steve Paulson on the *To the Best of Our Knowledge* radio show, May 26, 2013, http://www.ttbook.org/book/transcript/transcript-jim-fadiman-psychedelics.

cocktails of drugs: Thanks to the Erowid website for pointing out this discrepancy between the preliminary and final studies. "Our best interpretation of the protocol described in the first paper is that the subjects were given a dose of methamphetamine and a dose of chlordiazepoxide; then at some point a bit later they were given fairly low doses of a psychedelic (50 mcg of LSD, or either a 100 mg or a 200 mg dose of mescaline); then at mid-day they were given another dose of either methamphetamine, or chlordiazepoxide," wrote the editors of this website, which houses the Stolaroff archival documents. "The Curious Omission of Methamphetamine and Benzodiazepines from 1966 Article 'Psychedelic Agents in Creative Problem-Solving,'" Erowid.org, May 23, 2013, https://www.erowid.org/archive/stolaroff/stolaroff_collection_article3_creativity_1965.shtml#introduction.

133 *Picking up on Member H's fantasy:* Ibid.

James Russell: Brier Dudley, "Scientist's Invention Was Let Go for a Song," *Seattle Times,* November 29, 2004, http://old.seattletimes.com/html/businesstechnology/2002103322_cdman29.html.

illumination-phase experiment: "The Curious Omission of Methamphetamine and Benzodiazepines from 1966 Article 'Psychedelic Agents in Creative Problem-Solving.'"

134 *effects of LSD on volunteers:* "People's Brains Scanned While on LSD in Study in Cardiff," BBC News website, March 5, 2015, http://www.bbc.com/news/uk-wales-south-east-wales-31740491.

a barrage of thought experiments: Ben Bryant, "'The Greatest Opportunity We Have in Mental Health': Inside the British Study Where Volunteers Mainline LSD," *Vice News* (online magazine), March 5, 2015, https://news.vice.com/article/the-greatest-opportunity-we-have-in-mental-health-inside-the-british-study-where-volunteers-mainline-lsd.

"enhancing creative thinking": Robin Carhart-Harris, "Your Brain on Psilocybin: How Magic Mushrooms Expand Consciousness," *Slate,* July 4, 2014, http://www.slate.com/articles/health_and_science/medical_examiner/2014/07/psilocybin_in_brain_scans_magic_mushrooms_mimic_sleep_and_enhance_associations.html.

135 *for fifteen minutes:* Matthew Hutson, "People Prefer Electric Shocks to Being Alone with Their Thoughts," *Atlantic*, July 3, 2014, http://www.theatlantic.com/health/archive/2014/07/people-prefer-electric-shocks-to-being-alone-with-their-thoughts/373936/.

11. How to Time-Travel

138 *"tricorder":* Nathan Shedroff and Christopher Noessel, *Make It So: Interaction Design Lessons from Science Fiction* (Brooklyn, NY: Rosenfeld Media, 2012), Kindle edition, 274.

"what that thing might one day be": Brian David Johnson, *Science Fiction Prototyping: Designing the Future with Science Fiction* (San Rafael, CA: Morgan & Claypool Publishers, 2011), Kindle edition, 12.

139 *"Captain Nemo":* Genrich Altshuller, *Chance for an Adventure*, published in Russia in 1991, quoted in the online archives of the Altshuller Foundation, http://www.altshuller.ru/world/eng/interview1.asp.

black sailor's cap: Genrich Altshuller's navy picture, from the Altshuller archives, http://www.altshuller.ru/photo/photo03.asp.

fix-it man: Genrich Altshuller, "life story" interview with Igor Vertkin, Part 6, 1985–1986, translated for the author by Thomas Kitson in 2014. This comes from a series of Russian-language interviews on the Altshuller archive: http://altshuller.ru/interview/interview5-2.asp.

inspector in a patent office: Genrikh Saulovich Altshuller biography on the International TRIZ Association website, http://matriz.org/about-matriz/about-founder/.

"techniques for solving inventors' problems": Altshuller, "life story" interview.

140 *hidden in the patent system:* Ibid.

no phone: Victor Fey told me that Altshuller had no phone as late as the 1970s, so I assume that this was true throughout his early life as well.

read through forty thousand of them: Glen Mazur, "Theory of Inventive Problem Solving (TRIZ)," QFD Institute, February 26, 1996, http://www.mazur.net/triz/.

blueprints of the past: G. Altshuller, *And Suddenly the Inventor Appeared*, trans. Lev Shulyak (Worcester, MA: Technical Innovation Center, 1996), 58.

141 *"without a map":* This quote comes from Altshuller's first book, which has never been published in English (or, rather, I could find no English translation, despite laborious searches). The book, *How to Learn to Invent*, debuted in Russia in 1961. These passages were translated for me by Thomas Kitson. The Russian-language text was downloaded from http://www.altshuller.ru/triz/triz49.asp.

"I prefer prognostic science fiction": Alexey Golubev, "Communism in One Garage Laboratory: Invention and Amateur Engineering as a Utopian Project

of Soviet Technical Intelligentsia," presented at the Great Experiment Revisited: Soviet Science and Techno-utopianism conference, Princeton University, February 2012, http://www.academia.edu/2373474/Communism_in_one_Garage_Laboratory_Invention_and_Amateur_Engineering_as_a_Utopian_Project_of_Soviet_Technical_Intelligentsia.

objects out of powder. Genrich Altshuller and Valentina Zhuravleva, "The Donkey Axiom," reprinted in *Ballad of the Stars*, trans. Roger DeGaris (Worcester, MA: Technical Innovation Center, 2005), 63.

142 *"mercilessly throwing out billions of new machines"*: Ibid., 48–49.

"the French engineer Leboeux": Genrich Altshuller, *The Innovation Algorithm*, trans. Lev Shulyak (Worcester, MA: Technical Innovation Center, 2000), 241–42.

"road to 'crazy ideas'": Ibid., 243.

How to Learn to Invent: Leonid Lerner's biographical essay in *The Innovation Algorithm*, 308.

143 *"We would also discuss . . . history"*: Victor Fey, interview with the author, 2013.

144 *"'it' concept around the globe"*: Andy Raskin, "A Higher Plane of Problem-Solving," *CNN Money*, June 1, 2003, http://money.cnn.com/magazines/business2/business2_archive/2003/06/01/343390/.

145 *difficult to summarize:* If you crave a comprehensive introduction to Altshuller's method, I suggest that you track down his books *The Innovation Algorithm* and *Suddenly the Inventor Appeared: TRIZ, the Theory of Inventive Problem Solving.*

"shackled" by what we already know: Altshuller, *How to Learn to Invent*, trans. Thomas Kitson.

power-storage system: KTH The Royal Institute of Technology, "Battery Design Could Reduce Electric Car Weight," *ScienceDaily*, June 4, 2014, www.sciencedaily.com/releases/2014/06/140604093533.htm.

146 *"burden" on the car:* Emilia Lundgren, "Structural Batteries—A Unique Compromise for Lighter Electric Cars," Chalmers website, December 5, 2013, http://www.chalmers.se/en/departments/ap/news/Pages/Structural-batteries-%E2%80%93-a-unique-compromise-for-lighter-electric-cars.aspx.

The ideal machine is no machine: Altshuller, *Suddenly the Inventor Appeared*, 106.

"crowd of miniature dwarfs": Ibid., 100–101.

148 *"basic laws of technological development"*: G. S. Altshuller and R. B. Shapiro, "On the Psychology of Inventive Creativity," *Problems of Psychology*, no. 6 (1956): 37–49. Translated for the author by Thomas Kitson. Text accessed at http://www.altshuller.ru/triz/triz0.asp.

149 *president of the American Psychological Association:* Andrew L. Comrey, "Joy Paul Guilford: 1897–1987," National Academy of Sciences, 1993, http://www.nasonline.org/publications/biographical-memoirs/memoir-pdfs/guilford-joy-paul.pdf.

150 *"inventive coffee cups":* Steven Smith, interview and subsequent correspondence with the author, 2014–15.

the coffee-cup problem: David G. Jansson and Steven M. Smith, "Design Fixation," *Design Studies* 12, no. 1 (1991): 3–11, https://www.tamu.edu/faculty/stevesmith/SmithCreativity/Jansson&Smith1991.pdf.

"you can't un-know it": Smith, interview with the author.

151 *avoiding difficult thinking:* Steven M. Smith and Julie Linsey, "A Three-Pronged Approach for Overcoming Design Fixation," *Journal of Creative Behavior* 45, no. 2 (2011): 83–91, https://www.tamu.edu/faculty/stevesmith/SmithCreativity/SmithLinsey2011.pdf.

looked like a line of stagecoaches: William L. Breton, *Railroad Depot at Philadelphia*, June 6, 1832, Atwater Kent Museum of Philadelphia, retrieved from https://commons.wikimedia.org/wiki/File:Breton_Railroad_Depot_at_Philadelphia_1832.png.

"contextual baggage": Smith, interview with the author.

152 *he created a fake corporate memo:* With the help of his research assistant, who dreamed up many details of the fiction.

advertisements for the fictional product: Matthew R. Redmond, Michael D. Mumford, and Richard Teach, "Putting Creativity to Work: Effects of Leader Behavior on Subordinate Creativity," *Organizational Behavior and Human Decision Processes* 55, no. 1 (1993): 120–51, http://econpapers.repec.org/article/eeejobhdp/v_3a55_3ay_3a1993_3ai_3a1_3ap_3a120-151.htm.

"You've got to bear in mind": Michael Mumford, interviews and subsequent correspondence with the author, 2014–15.

153 *The judges were so impressed that:* Ibid.

Memex in his imagination: Matt Mahoney, "Future Perfect," *MIT Technology Review,* October 27, 2010, http://www.technologyreview.com/fromthearchives/421424/future-perfect/.

"barred to the machine": Vannevar Bush, "Memex II," essay discovered among his papers and dated August 27, 1959. Reprinted in *From Memex to Hypertext*, ed. Nyce and Kayn, 183.

12. The Go-Betweens

157 *ships crashed:* James Herbert Cooke, "The Shipwreck of Sir Cloudesley Shovell, on the Scilly Islands in 1707," read at a meeting of the Society of Antiquaries, London, February 1, 1883, retrieved from http://www.hmssurprise.org/shipwreck-sir-cloudesley-shovell.

158 *to collect celestial observations:* Richard Dunn, "The History," Longitude Prize website, https://longitudeprize.org/history.

159 *"conjure up a crowd":* Karim Lakhani, interview with the author in 2014 and subsequent correspondence in 2015.

"eBay of innovation": Dwayne Spradlin, interview with Molly Petrilla for *ZD-*

Net, August 6, 2012, http://www.zdnet.com/article/qa-dwayne-spradlin-on-leading-a-marketplace-of-innovation-innocentive/.

to answer their questions: Dave Aron, "InnoCentive Positions Itself as a Challenge Marketplace," Gartner, Inc., June 27, 2012, https://www.innocentive.com/files/node/casestudy/gartner-report-innocentive-positions-itself-challenge marketplace.pdf.

three hundred thousand "solvers": "InnoCentive Solver Network Passes 300,000 Registered Members Milestone," InnoCentive website, August 19, 2013, https://www.innocentive.com/innocentive-solver-network-passes-300000-registered-members-milestone.

"fractal expressionist": "Solver Profile," InnoCentive website, https://www.innocentive.com/ar/solver/show/324682.

160 *"John Harrison was a nobody":* Lakhani, interview and correspondence with the author, 2014–15.

data generated by: Lars Bo Jeppesen and Karim R. Lakhani, "Marginality and Problem-Solving Effectiveness in Broadcast Search," *Organization Science* 21, no. 5 (2010): 1016–33, http://dl.acm.org/citation.cfm?id=1862443.

"unwanted food coloration": "Heat Stable Prevention of Flavan3-ols–Iron (II) Interactions," September 5, 2012, InnoCentive website, https://www.innocentive.com/ar/challenge/9932981.

161 *"no way on Earth":* Dwayne Spradlin, interview and correspondence with the author, 2014–15.

162 *"some kid in Estonia":* Lakhani, interview with the author, 2014.

whether women were more likely than men: Jeppesen and Lakhani, "Marginality and Problem Solving Effectiveness in Broadcast Search."

163 *"A Diabetic Cookie":* "A Diabetic Cookie That Tastes Like the 'Real Thing,'" InnoCentive website, November 16, 2012, https://www.innocentive.com/ar/challenge/9933136.

"I made the cookies myself": Vaishali Agte, interview and subsequent correspondence with the author, 2014–15.

164 *$5,000 for the right to use her recipe:* "A Diabetic Cookie That Tastes Like the 'Real Thing.'"

no pay at all: Agte, interview with the author.

a solution they already knew about: Jeppesen and Lakhani, "Marginality and Problem-Solving Effectiveness in Broadcast Search."

challenge in cosmetic dentistry: Cristina Jimenez, "Funding: Researching Outside the Box," *Nature, Nature Jobs* (blog), January 11, 2011, http://www.nature.com/naturejobs/science/articles/10.1038/nj7330-433a.

165 *"wild and crazy things":* Leslie Koren, "Self-Tanning Is in High Demand," *The Free Lance–Star* (Fredericksburg, VA), June 15, 2003.

166 *Oompa Loompa:* Tom Laughlin, interviews and correspondence with the author, 2012–15.

"the concept of spraying a person in a booth": Ibid. Also see Thomas Laughlin, U.S. Patent 199557—"Method of and Apparatus for Automatically Coating the Human Body," accepted on March 13, 2001.

13. Zones of Permission

168 *"Every active and ingenious farmer":* "Brief and Seasonable Hints," *The Cultivator & Country Gentleman,* March 10, 1870, 147.

169 *"failed ten-thousand times":* The origin and precise formulation of this quote are subject to some dispute—see "I Have Gotten a Lot of Results! I Know Several Thousand Things That Won't Work," *Quote Investigator* (blog), July 31, 2012, http://quoteinvestigator.com/2012/07/31/edison-lot-results/#more-4181.

slab covered in glass: Michael Calafati et al., eds., *The Bell Labs Charrette: A Sustainable Future* (New Jersey Society of Architects, 2008), http://docomomo-us.org/files/bell_labs_final%20book.pdf.

chance encounters: Kim Steele, "Bell Labs—Our Scientific Heritage," *The Boulevardiers* (blog), November 11, 2013, http://theboulevardiers.com/2013/11/11/bell-labs-our-scientific-heritage/.

170 *first computerized card catalog:* A. Michael Noll, "You Had to Be There: Bell Telephone Laboratories in the 1960s," personal remembrance, April 8, 2012, http://noll.uscannenberg.org/BellLabs.html.

five thousand employees: Jon Gertner, *The Idea Factory: Bell Labs and the Great Age of American Innovation* (New York: Penguin Group, 2012), Kindle edition, 284.

"problem-rich environment": Ibid., 47.

"led to good inventions": Ibid., 49.

171 *"skeletons in the imperial closet":* Wu, *The Master Switch,* Kindle locations 1891–1994.

172 *"no matter how far afield":* Griffin et al., *Serial Innovators,* Kindle locations 2005–9.

"put their jobs on the line": Ibid., Kindle locations 1927–28.

if he could stack up layers: Chuck Hull, interview with the author, 2013. See also Pagan Kennedy, "Who Made That 3-D Printer?" *New York Times Magazine,* November 22, 2013, http://www.nytimes.com/2013/11/24/magazine/who-made-that-3-d-printer.html?_r=0.

173 *"killing the US automotive industry":* Hull, interview with the author.

proposed the 3-D printer: Note that the term *3-D printer* was not yet in use. Hull called his invention "stereolithography." However, I have used the term *3-D printing* here to avoid confusion.

"kludged together": Kennedy, "Who Made That 3-D Printer?"

175 *filled a glue gun:* Bien Perez, "3D Printing Pioneer Scott Crump's Kitchen Experiment," *South China Morning Post,* July 22, 2013, http://www.scmp.com/business/companies/article/1287961/3d-printing-pioneer-scott-crumps-kitchen-experiment.

"voted with their feet": Fred Block and Matthew R. Keller, *Where Do Innovations Come From? Transformations in the U.S. National Innovation System, 1970–2006*, report issued by the Information Technology & Innovation Foundation, July 2008, http://www.itif.org/files/Where_do_innovations_come_from.pdf.

176 *disproportionate share of the most valuable ideas:* Sadao Nagaoka and John P. Walsh, "The R&D Process in the U.S. and Japan: Major Findings from the RIETI-Georgia Tech Inventor Survey," working paper from the Research Institute of Economy, Trade and Industry, July 5, 2009, 1, http://www.rieti.go.jp/jp/publications/dp/09e010.pdf.

forty-seven years old: Ibid., 10.

177 *"the corporation is founded on boxes":* Cooper, interview with the author.

"tools for creation are becoming democratized": Lakhani, interview with the author.

178 *she "embedded" herself at the space agency:* Hila Lifshitz-Assaf, "From Problem Solvers to Solution Seekers: Dismantling Knowledge Boundaries at NASA," a working paper submitted for review at the Harvard Business School, May 1, 2014. Because this paper had not been formally published, I corresponded with Lifshitz-Assaf to confirm the facts and quotes in it.

NASA had been searching: Ibid., 18.

NASA offered a prize: NASA partnered with InnoCentive to design and administer the problem-solving challenge.

75 percent accuracy: "NASA Announces Winners of Space Life Sciences Open Innovation Competition," NASA, July 7, 2010, http://www.innocentive.com/nasa-announces-winners-space-life-sciences-open-innovation-competition-0.

NASA was "stunned": Lifshitz-Assaf, "From Problem Solvers to Solution Seekers," 19.

179 *"shame was palpable":* Ibid., 22.

"'What value am I?'": Ibid., 43.

"invent-it-ourselves model": Ibid., 37.

14. Holistic Invention

180 *one out of every six deaths:* John T. James, "A New, Evidence-Based Estimate of Patient Harms Associated with Hospital Care," *Journal of Patient Safety* 9, no. 3 (2013): 122–28, http://journals.lww.com/journalpatientsafety/Fulltext/2013/09000/A_New,_Evidence_based_Estimate_of_Patient_Harms.2.aspx.

181 *seventy spinoff companies:* "About Us," Cleveland Clinic Innovations website, http://innovations.clevelandclinic.org/About-Us/Innovations-Team.aspx.

artificial heart: Ibid. See also Monica Robins, "Cleveland Clinic Invents Artificial Smart Heart," WKYC.com, September 11, 2012, http://www.wkyc.com/news/article/258085/3/Cleveland-Clinic-invents-artificial-smart-heart.

Dr. Yogen Saunthararajah: Astute readers will notice that we first encountered Sauntharajah in Chapter 6, where he weighed in on the challenges of developing cancer therapies. The quotes from Sauntharajah in this chapter also come from my interviews and correspondence with him in 2014 and 2015.

182 *three million Americans: Preventing Central Line–Associated Bloodstream Infections: A Global Challenge, a Global Perspective*, a report issued by a consortium of medical organizations in 2012, v, http://www.jointcommission.org/assets/1/18/CLABSI_Monograph.pdf.

About thirty thousand Americans die: Ibid., x. Also see "Hospitals Profit When Patients Develop Bloodstream Infections," *ScienceDaily*, May 22, 2013, a summary of a Johns Hopkins University report, www.sciencedaily.com/releases/2013/05/130522141841.htm.

nine days inside the body: L. J. Worth and M. L. McLaws, "Is It Possible to Achieve a Target of Zero Central Line Associated Bloodstream Infections?" *Current Opinion in Infectious Disease* 25, no. 6 (2012): 650–57, http://www.ncbi.nlm.nih.gov/pubmed/23041775.

"part that hangs outside the body": Catherine Musemeche, correspondence with the author, 2014.

"Teeth That Think": Clay Risen, "Teeth That Think," *New York Times Magazine*, June 3, 2012, http://query.nytimes.com/gst/fullpage.html?res=9C0DE1D61531F930A35755C0A9649D8B63. See also Neil Versel, "Prototype iPhone Biosensor Detects Viruses, Bacteria, Toxins, Allergens," *Mobi Health News*, May 30, 2013, http://mobihealthnews.com/22691/prototype-iphone-biosensor-detects-viruses-bacteria-toxins-allergens/. Further note: Sauntharajah couldn't remember exactly which article inspired his revelation, so I scoured the *New York Times* archives and found the only story from that year that involved medical biosensors. I think it's reasonable to assume that the article named here was the one that inspired his idea.

183 *"gave me a window":* Sauntharajah, interview with the author.

184 *"could be really important":* Christine Moravec, interview and follow-up correspondence with the author, 2014.

"during the first year [2011]": "InnoCentive and Cleveland Clinic Collaborate to Solve Healthcare's Biggest Challenges," InnoCentive, Inc., July 12, 2011, http://www.innocentive.com/innocentive-and-cleveland-clinic-collaborate-solve-healthcare%E2%80%99s-biggest-challenges.

"Cleveland Clinic has to invest millions of dollars": Moravec, interview with the author.

185 *"become contaminated":* "Cleveland Clinic: Early Warning Indicator for Contamination or Fouling of a Central Venous Catheter," InnoCentive challenge announcement, https://www.innocentive.com/ar/challenge/9933157.

team of Spanish researchers: J. Paredes et al., "Smart Central Venous Port for Early Detection of Bacterial Biofilm Related Infections," *Biomedical Microdevices* 16, no. 3 (2014): 365–74, http://www.ncbi.nlm.nih.gov/pubmed/24515846.

186 *"I didn't have any credibility":* Christie Riggins, "Great Inventions Don't Happen Overnight," *Stanford Medicine* 17, no. 3 (2000), http://sm.stanford.edu/archive/stanmed/2000fall/inventions.html.

"two rarely meet": Stephen C. Schimpff, "The Future of Medicine," in *The Future of Medicine* (Nashville: Thomas Nelson, 2007), 118.

other top hospitals: Moravec, interview with the author. See also David Cameron, "The Ideation Challenge on Diabetes," Harvard Medical School website, http://hms.harvard.edu/news/harvard-medicine/ideation-challenge-diabetes.

187 *Photographs from the 1930s:* I researched this paragraph by studying numerous photos of 1930s car wrecks posted on Flickr.

A few doctors became so concerned: "Watertown Man Invented Seat Belt System in 1950s," *Aberdeen (SD) News,* September 2, 2007, article collected on *American News* and accessed here: http://articles.aberdeennews.com/2007-09-02/news/26400010_1_seat-belts-family-car-family-practice. See also Lee Vinsel, "Doctors Inventing Auto Safety," *Bright Ideas* (blog affiliated with the Smithsonian's Lemelson Center for the Study of Invention and Innovation), August 2, 2013, http://blog.invention.smithsonian.org/2013/08/02/doctors-inventing-auto-safety-2/.

"injury-producing accident": Ralph Nader, *Unsafe at Any Speed: The Designed-In Dangers of the American Automobile* (New York: Grossman Publishers, 1965), accessed online from http://www.american-buddha.com/nader.unsafeanyspeed.5.htm.

188 *"As soon as the youngsters":* Ibid.

15. Paper Eyes

193 *"perseverance":* Joseph Rossman, *Industrial Creativity: The Psychology of the Inventor* (New Hyde Park, NY: University Books, 1964), 40.

194 *received his first patent:* It's hard to establish an exact date when Altshuller filed his first patent. Since accounts vary, I drew from Leonid Lerner's account in *The Innovation Algorithm.*

premier seed bank: Charles Siebert, "Food Ark," *National Geographic,* July 2011, http://ngm.nationalgeographic.com/2011/07/food-ark/siebert-text/2.

195 *Soviet patent rate:* It's very difficult to confirm the actual patent rate at that time, so here I am depending on Altshuller's account of the numbers.

"How could we talk about inventing?": Genrich Altshuller, "life story" interview with Igor Vertkin, Part Nine, translated for the author by Nataliya Skoryk, accessed in Russian from http://www.altshuller.ru/interview/interview5-4.asp.

letter to Stalin: It's hard to know exactly what was in this letter. Though Altshuller described it later, I could find no existing copies of the document.

sent their letter: Altshuller, "life story" interview, Part Nine.

the military police ambushed: Lerner, biographical sketch in *The Innovation Al-*

gorithm. This account also comes from my immersion in numerous other accounts of Altshuller's life.

196 *dragged him to a prison:* The account of Altshuller's travails in prison, including his interrogation, are drawn from many sources. They include the following: Lerner's biographical sketch in *The Innovation Algorithm;* Y. B. Karasik, "Towards Altshuller's 80th Birthday," an account of the author's memories of Altshuller, November 2005, http://www3.sympatico.ca/karasik/towards_altshuller_80_birthday.html; Maharram Zeynalov, "Universal Invention," *Region Plus* (magazine about Azerbaijan), October 10, 2013, http://www.regionplus.az/en/articles/view/1628; and Genrich Altshuller, "The Technology of Memoir Writing," 1986, translated for the author by Thomas Kitson, accessed in Russian from http://www.altshuller.ru/stories/story4.asp.

197 *to make those eyes:* Quoted in Zeynalov, "Universal Invention."

200 *"honor, glory, valor":* "Gulag: Many Days, Many Lives," Roy Rosenzweig Center for History and New Media, 2008–2015, Item #291, http://gulaghistory.org/items/show/291.

201 *"Before prison":* Lerner, biographical sketch in *The Innovation Algorithm.*

flecked with bone: Ibid. I also drew on my deep reading of multiple sources to weave together the details in this section.

202 *"There was strong resistance":* G. S. Altshuller, "The History of ARIZ Development," originally presented at the "Theory of Inventive Problem Solving" seminar at the School of Management, Simferopol, Ukraine, 1986, retrieved from http://www.ideationtriz.com/paper_ARIZontheMove.asp#The_History_of_ARIZ_Development.

203 *"a swarm of flies or just a cloud":* "Genrich Altshuller Teaching TRIZ 5 of 6," YouTube, a clip from a documentary shot in Buku in the 1970s, uploaded April 17, 2007, https://www.youtube.com/watch?v=lsQ6cRzrlV0.

shut it down: Altshuller biography, the International TRIZ Association website, http://matriz.org/about-matriz/about-founder/.

"creative individuals in the population": Genrich Altshuller and Igor Vertkin, *How to Become a Genius,* published in Russian in 1994. This quote was accessed from http://www.altshuller.ru/world/eng/interview1.asp.

"crazy inventors": Altshuller, *The Innovation Algorithm,* 18.

204 *TV show pitted kids against professional engineers:* Boris Zlotin et al., "TRIZ Beyond Technology: The Theory and Practice of Applying TRIZ to Non-Technical Areas," Ideation International Inc., February 2000, http://www.ideationtriz.com/paper_TRIZ_Beyond_Technology.asp#Development of the creative personality#Development of the creative personality.

never seemed interested in making money: Isak Bukhman, *TRIZ—Technology for Innovation* (Taipei: Cubic Creativity Company, 2012), 13. Bukhman reports that during his intensive studies with Altshuller, his teacher never charged tuition or any kind of fee.

205 *Dow Chemical to Ford to Samsung:* "Mumbo Jumbo?" Institution of Mechani-

cal Engineers website, 2004, http://www.imeche.org/knowledge/industries/manufacturing/triz/mumbo-jumbo.

shave production costs: Steve Hamm, "Tech Innovations for Tough Times," Bloomberg Business website, December 25, 2008, http://www.bloomberg.com/bw/stories/2008-12-25/tech-innovations-for-tough-timesbusinessweek business-news-stock-market-and-financial-advice.

"without threatening the stability of the company": Subir Chowdhury, *Design for Six Sigma* (Chicago: Dearborn, 2002), 114.

"Don Quixote's lunge": G. Altov (a.k.a. Genrich Altshuller), *How to Become a Heretic*, published in Russian in 1991, translated by Google. I accessed the text from http://www.altshuller.ru/world/eng/interview1.asp.

16. Tinkering with Education

206 *"I'll do anything to get in":* Neil Gershenfeld, *Fab: The Coming Revolution on Your Desktop—From Personal Computers to Personal Fabrication* (New York: Basic Books, 2008), Kindle edition, 107–19.

could cost $50,000: Rachael King, "Printing in 3D Gets Practical," Bloomberg Business website, October 6, 2008, http://www.bloomberg.com/bw/stories/2008-10-06/printing-in-3d-gets-practicalbusinessweek-business-news-stock-market-and-financial-advice.

207 *a billion people had no access:* "Low-Cost Prescription Eyeglass Lens Fabricator, Saul Griffith, 2001–2004," MIT Museum, 2011, http://bit.ly/1G9aTpH.

bandage dangling from one hand: The details and quotes in this section are drawn from the author's interview with Saul Griffith in 2004, and from follow-up communications with Griffith in 2014 and 2015.

209 *"solving the wrong problem":* David Owen, "The Inventor's Dilemma," *The New Yorker*, May 17, 2010, http://www.newyorker.com/magazine/2010/05/17/the-inventors-dilemma.

210 The Boy Mechanic: A copy of the book (complete with the illustration) is available at Project Gutenberg, http://www.gutenberg.org/files/12655/12655-pdf.pdf.

211 *"So what kind of machine do you want to invent tonight?":* The details about the Howtoons Club come from my reporting and tape recordings made in June 2004.

212 *"end up as auto mechanics":* Ben Trettel, interview with the author, 2013, and follow-up correspondence in 2015.

"Because of the water guns": Ibid. Also see the Trettel interview in Kennedy, "Who Made That Super Soaker?"

213 *"Spatial ability is a powerful":* David Lubinski, "Spatial Ability and STEM: A Sleeping Giant for Talent Identification and Development," *Personality and Individual Differences* 49, no. 4 (2010): 344–51.

spatial-reasoning tests could be used: Harrison J. Kell et al., "Creativity and Technical Innovation: Spatial Ability's Unique Role," *Psychological Science* 24, no. 9

(2013): 1831–36, https://my.vanderbilt.edu/smpy/files/2013/02/Kell-et-al.-2013b.pdf.

"I run my turbine in thought": Tesla, *My Inventions*, 13.

spatial reasoning can be learned: David H. Uttal et al., "The Malleability of Spatial Skills: A Meta-Analysis of Training Studies," *Psychological Bulletin* 139, no. 2 (2013): 352–402, http://longevity3.stanford.edu/brainhealth/files/2013/04/uttal-et-al2013_malleability-of-spatial-skills-meta-analysis.pdf.

214 *More than five thousand high schools:* "MakerBot Partners with D&H to Distribute 3D Printers and Scanners in Education, Healthcare and Government Verticals," Stratasys press release, April 15, 2015, http://investors.stratasys.com/releasedetail.cfm?ReleaseID=906719.

organizations including the Department of Agriculture: "Fact Sheet: President Obama to Host First-Ever White House Maker Faire," White House, Office of the Press Secretary, June 18, 2014, https://www.whitehouse.gov/the-press-office/2014/06/18/fact-sheet-president-obama-host-first-ever-white-house-maker-faire.

considered eight thousand proteins: Jack Andraka and Matthew Lysiak, *Breakthrough: How One Teen Innovator Is Changing the World* (New York: HarperCollins, 2015), Kindle edition, 100.

215 *using the Internet, Andraka:* Ibid., 132.

"prodigy of pancreatic cancer": Abigail Tucker, "Jack Andraka, the Teen Prodigy of Pancreatic Cancer," *Smithsonian* magazine, December 2012, http://www.smithsonianmag.com/science-nature/jack-andraka-the-teen-prodigy-of-pancreatic-cancer-135925809/.

"like a powerful weapon": Andraka and Lysiak, *Breakthrough*, 42.

Conclusion

216 *"not evenly distributed":* "The Science in Science Fiction," interview with Gibson on the NPR show *Talk of the Nation*, November 30, 1999, http://n.pr/1dASzLa.

217 *"successive layers of Tomorrowlands":* Pagan Kennedy, "William Gibson's Future Is Now," *New York Times Book Review*, January 13, 2012, http://www.nytimes.com/2012/01/15/books/review/distrust-that-particular-flavor-by-william-gibson-book-review.html.

toothbrush originated in China: "Prison, Suicide, & the Cold-Climate Hog (the Sordid History of the Toothbrush)," description of an exhibition at the Museum of Everyday Life, no date given, http://museumofeverydaylife.org/exhibitions-collections/previous-exhibitions/toothbrush-from-twig-to-bristle-in-all-its-expedient-beauty/a-visual-history-of-the-toothbrush.

218 *"Can't we do better?":* Sauntharajah, interview with the author.

"yin and yang of modern science": Roberta Ness, *The Creativity Crisis: Reinventing Science to Unleash Possibility* (Oxford: Oxford University Press, 2014), Kindle edition, 5.

219 *strong bones and joints:* H. Chirchir et al., "Recent Origin of Low Trabecu-

lar Bone Density in Modern Humans," *Proceedings of the National Academy of Sciences* 112, no. 2 (2015): 366–71, http://www.researchgate.net/profile/Brian_Richmond/publication/270571532_Recent_origin_of_low_trabecular_bone_density_in_modern_humans/links/54ada8950cf2828b29fcb19c.pdf.

"People were adopting farming": Helen Thompson, "Switching to Farming Made Human Joint Bones Lighter," Smithsonian.com, December 22, 2014, http://www.smithsonianmag.com/science-nature/switching-farming-made-human-joint-bones-lighter-180953711/#TH0Oo2zSiXcvlqFm.99.

220 *"That's what democratization really does":* Butte, interview with the author.

concierge for inventors: Tully Gehan, interview with the author, 2014.

221 *"My job used to be so hard":* Hawker, interview with the author.

222 *more than four thousand backers:* "Carbon Flyer: The Ultimate Crash Proof Video Drone," Ingiegogo website, https://www.indiegogo.com/projects/carbon-flyer-the-ultimate-crash-proof-video-drone#/updates.

INDEX

Abba, Mohammed Bah, 28
accidental discovery
  "aha" moment of, 52,
  being open to, 55–58, 68
  as byproduct of other research, 48–50, 61–62, 63
  engineering, with data mining, 70–73
  hidden usefulness of, not immediately clear, 49–50, 62
  luck and, 51–53, 54, 55–58
  overview, 48, 49
  perception and, 53, 54
  serendipity and, xvi, 47–49, 53–56, 64, 68
  Super-Encounterers and, 55–56, 68, 76
  tinkering leading to, 59–60
  and zones of permission, 172
Adhesive Technologies, 31–34
Advanced Tissue Services, 75
aeronautics industry, 93
agriculture, 85, 92, 93, 194
Agte, Vaishali, 163–64
"aha" moment, 52, 61, 112
Airbnb, 75
air quality
  smoke detector, 49–51
  testing, 38–39
Albright, Richard, 121–22
Alcorn, Allan, 106–7
Alibaba, 221
Alternative Economy Revolution, 21
Altshuller, Genrich, xii–xiii, 138–49, 194–205, 213, 252n
aluminum, 93
Amazing Stories (magazine), 137
Amazon, xiv
Ames Research Center, 111
Amgen, 66
Amoco, 144, 205
Ampex, 125, 126, 130–31
amphetamines, 132, 243n
Anchor, 17, 18
Andersen, Buzz, 10
Andersen, Robert, 10
Anderson, Tim, 87–91, 173
Andraka, Jack, 214–15
animation, 123
Apple, 177, 183
apps
  Apple's creative marketplace for, 177
  Birdi air quality monitoring, 38
  Mars-time alarm clock, 7–8
art
  creative process, 56–58, 153, 165
  drawings, role in invention, xi, 57–58, 84, 110, 139, 173
  fashion and color trends, 81
  strobe photography, 59
Artisan's Asylum, 91

aspartame, 48–49
Assay Depot, 74
AT&T, 101–4, 170–71, 238–39n
Auyang, Sunny, 49, 53
Azerbaijan, xii, 139. *See also* Soviet Union
Azerbaijan Public Institute of Inventive Creativity, 202–3

Baker, Ted, 85
Bass, Deborah, 7
Belanger, Bryan, 32, 33
Belanger, Dick, 31–34, 220
Belanger, Steve, 33
Belinsky, Mark, 24, 38–39
Bell Labs, 101, 102, 106, 169–71
benzene, xv
bioinformatics (biology *in silico;* dry biology), 64–65, 69–70
biotechnology, 61, 74–76, 183–85, 220
Birdi, 38–39
Bonsen, Joost, 210
Bosch, Carl, 92, 93, 238n
Botswana, 26–27
Bradin, David, 160
Brand, Stewart, 131, 204
Breakthrough Award, 29
Bredt, Jim, 89–92, 173
Building 20, 87–92, 175
Burke, Colin, 109
Burnham, Scott, 60
Bush, Vannevar, 108–11, 112, 153
Bushnell, Nolan, 106–7
business machines
  copy machine, 84
  fax machine, 6, 171
  See also printers
Butte, Atul, 70–73, 74–75, 220

cameras
  to photograph thoughts, 128–29
  video, 51–52
Candy Machine, 90
Carbon Flyer toy, 222
Cardiff University, 134
Carhart-Harris, Robin, 134
Carlson, Chester, 84
cars, 9, 145, 151–52, 166, 174, 187–88
cell phones
  Apple iPhone, 177, 183
  car-based, 101–2, 239n
  developed and scuttled, 171
  hand-held, xv–xvi, 102–5, 117–18, 217, 238–39n
  infrastructure to support, 102, 104
  from multiple breakthroughs, 97
  pager as precursor to, 98–100
  in science fiction, 99, 104–5
  text and Twitter on, 8–10
chemistry
  artificial sweeteners, 48–49
  artificial tanning ingredients, 165–66
  benzene molecular structure, xv
  diabetic cookie development, 163–64
  gender bias in the discipline, 162–63
  health shake development, 160–61, 165
China, 74, 217, 220–21
chips
  gene, DNA microscopy on, 70–71
  silicon, 119, 120–21 (*see also* bioinformatics)
Chirchir, Habiba, 219
chlordiazepoxide, 243n
Chomsky, Noam, 88
Chowdhury, Subir, 205
civic engagement, 187–89, 201, 203, 208
Cleveland Clinic, 181–87
closed system, 171, 175, 177
collaboration
  advantages of, 158, 218, 219–20
  through crowdfunding, 36–42
  in zones of permission, xi, 169–70, 177
color trends, 81
comics
  Dick Tracy, 99, 104
  Howtoons, 210–11

communication
collaboration, xi, 36–42, 158, 169–70, 177, 218, 219–20
with customers, to define and fill a need, 30
doctor-patient, 64, 76
emoticons, 5
symbols as shortcuts for typing, 10–11
See also feedback
community of ideas
importance of, 218
prosthetic devices, 17–20, 230n
social media, 165
through crowdfunding, 36–42
through InnoCentive, 159–64, 165, 184–85, 186
through the Internet, 14, 19, 21, 177, 179
by user-inventors, 11, 12–14, 17–20, 230n
competitions
Breakthrough Award, 29
InnoCentive, 159–64, 165, 184–85, 186
Intel ISEF Gordon E. Moore Award, 215
Lemelson-MIT National Collegiate Student Prize, 27, 209
NASA, solar-particle storm prediction, 178–79
problem-solving, 159–60
R&D 100 Awards, 175–76
computers
Altair 8800, 118
Alto, 116–18
CALDIC, 112–13
ENIAC, 109
Memex, as conceptual precursor, 108–11, 112, 153
personal, Engelbart's vision and demo, 112–15
personal, Xerox development, 115–17
computer software
animation, 123
data-mining techniques, 64–65
dispatch, 9
training in Altshuller's theories, 144
UNIX, 169
computer technology
integrated circuit, 119, 120–21, 146
Moore's Law, 120–21, 122–23, 130
mouse, xv, 114
silicon chip, 119, 120–21 (*see also* bioinformatics)
connecting
Go-Betweens, 157–67
holistic invention, 180–89
overview, xvii
zones of permission, 168–79
See also collaboration; communication; community of ideas
contradiction paradox, 145–46, 202
Coolest Cooler, 40
Cooper, Martin, xv–xvi, 30, 97–98, 100–105, 117, 176–77
copy machine, 84
Cote, Fred, 88
Cragin, Bruce, 178
creativity
"aha" moment, as overrated, 52, 61, 112
competitions, problem-solving (*see* competitions)
LSD to enhance, 126–27, 130–34, 243n
"luck" as a form of, 51–53
mental blocks to, 144, 149–52, 164–65, 193
merging of diverse disciplines, xvii, 158–61, 164–65, 166, 172, 185–86, 219–20
micro-creativity, xii
mindset during problem-solving, 51, 86, 91, 116, 145–46, 153, 202
mindset during research, 54–56
notebook of ideas/sketches, 31, 33, 61

process of imagination (*see* imagination, psychology of; thought process, mind experiments)
role of play, 51, 57, 59, 89, 91, 169
spatial reasoning as predictor of, 213
zones of permission, 168–79
criticism. *See* feedback
cross-pollination. *See* creativity, merging of diverse disciplines
crowdfunding
crowd-whispering, 37–40
failure on, 39–42
Indiegogo, 36, 39, 40, 222
Kickstarter, 36, 37–38, 39–40, 221, 232n
overview, xiii
as pre-mortem, 36–40
Crump, S. Scott, 175
Csikszentmihalyi, Mihaly, 56–58
Culkin, John, 106
cups
sippy, 32–33, 220
for takeout coffee, 149–51
Cycle Tech, 86

dark matter, 14–15, 177. *See also* user-inventors
data algorithms, 69–70, 71, 72
data mining
to analyze customer preferences, 71
bioinformatics, 64–65, 69–70
to discover the principles of creativity, 140, 160, 195
for drug development, 64–65, 70–73, 77
social media information, 165
to track color trends, 81
David, Larry, 79
da Vinci, Leonardo, 110
Defense Advanced Research Projects Agency (DARPA), 16, 214
déformation professionnelle, 164–65
Dennison, Greg, 20
Derek (child), 20
Derk, Tim, 12–13
developing countries
eyeglass lens fabricator, 207–9
low-cost technology for, 23–29
malaria and starvation in, 111
Dickman, Steve, 65
Dick Tracy (comic strip), 99, 104
The Difference Engine (Gibson and Sterling), 216–17
discovery
building an empire out of nothing, 79–94
as byproduct of other research, 48–50, 61–62, 63
data goggles, 63–78
easy in hindsight, x
overview, xvi
Super-Encounterers, 47–62, 68, 76
through serendipity, xvi, 47–49, 53–56, 64, 68
discrimination, 161–63
distribution
Alternative Economy Revolution, 21
direct sales through cottage industry from home, 33
direct sales through the Internet, 32, 221–22
relationship with local infrastructure, 209
supply chain, xi
through mass marketing company, 42
Djerassi, Carl, 93
"The Donkey Axiom" (Altshuller), 141, 146
Dorsey, Jack, 8–10
Dow Chemical, 205
Dragotta, Nick, 210
drawing, xi, 57–58, 84, 110, 139, 173
Drouillard, Jamy, 25–26

drug development
big pharma vs. little pharma, 73–78
costs, 62, 65, 77
data mining for, 64–65, 71–73, 77
from plants, 93
targeted searches, 63–64
traditional trial-and-error hypothesis-testing, 77
drugs to enhance the imagination, 125–27, 130–34

early adopters, 99, 118
Edgerton, Doc, 59–60
Edison, Thomas, xiv, 168–69
education
in creative problem solving, 143–49, 195, 201–5
empathy in classes, 24
Howtoons, 210–11
limitations of current, 212
national agencies/organizations with programs, 214
in spatial reasoning, 213
teach children to challenge the status quo, xvii, 210
in tinkering, 24, 209–15
Eindhoven, Netherlands, xii
Einstein, Albert, 128
empathy, 23–30
empowerment
overview, xvii, 193–94
perseverance and creativity, 193–205
through education (*see* education)
e-NABLE, 19–20
energy production
battery power for electric car, 145
charcoal fuel from sugar cane, 28–29
methane, 85–86
printer to generate solar panels, 37
solar cells, 169
Windbelt generator, 29
Engelbart, Christina, 112
Engelbart, Doug, xv, 111–15, 119, 131, 204
engineering
backward, 111
Edison's idea factory, 168–69
fabrication, 173–74, 206–7
gender bias, 162–63
Howtoons projects, 210–11
of prosthetic devices, 16–20, 230n
role in invention, xi, 23–29, 111, 146
England, 158, 159
English, Bill, xv
Erdelez, Sanda, 54–55
Errami, Mounir, 164
ethical issues, 42
ethnography, 23–30, 101
evolution
human, 219
of the invention process, 222–23
experience, invention of an, 98, 107, 114, 117–18
extrapolation, in futuristic thinking, 110–11, 112, 118, 123

fabrication, 173–74, 206–7
Facebook, 71, 175, 177
Factory For All, 220–21
Fadiman, Jim, 132
failure
ability to imagine, 34–35, 36, 61
on crowdfunding, as pre-mortem, 39–42
using, to then discover product, 59–60, 169
Fairchild Semiconductor, 119
fashion, 81
fax machine, 6, 171
Federal Communications Commission (FCC), 101–2, 104
feedback
criticism and the ability to adapt, 30, 34–35, 40
the future of, 31–43
"I could have thought of that," x

ignored due to inventor's over-investment, 35
impossibility of invention, xvi, 113, 114
and institutional status quo, 171, 172
Lead Users have most valuable, 41, 118, 181
learning from (pre-mortem), 34–42
limitations of surveys, 41
online reviews, xiv
opposition to invention, xvii, 35, 115, 130, 171–74, 179
rejection of inventor, xvii, 84, 113–15, 162, 172, 179, 186
self-criticism, 193
through crowdfunding response, 37–40
usefulness/uselessness, 30, 84–85
fertilizer, 92, 93
Fey, Victor, 143–44, 203
fiber optics, 62, 171
fieldwork, importance of, 23–30, 85–86, 101, 172, 178–79, 208
films
animation, 123
Netflix, 71
Toy Story, 123
Wall Street, 104
financial aspects
drug development, 16, 62, 74, 77
low-cost technology for developing countries, 24, 27, 29, 209
products that require expensive R&D, 15
prosthetic devices, 18, 20, 231n
sponsorship through a company, 32, 41–42
sponsorship through the Internet (*see* crowdfunding)
fire alarms, 50
Flaherty, Alice, 82–85
Fleming, Alexander, xvi, 80
Fogarty, Thomas, 185–86
food and food science
artificial sweeteners, 48–49, 90–91
colorants, 81
color removal, 160–61
crop fertilizer, 92, 93
diabetic cookie, 163–64
grain mill, motorized, 27
health shake, 160–61, 165
heating for hydroponic greenhouse and fish farm, 85
secure food system, 92, 93, 194
Ford, 205
Ford, Henry, 41, 232n
Frayne, Shawn, 28–29, 30, 36–37, 41, 73, 232n
Fry, Art, xi
Fuller, Buckminster, 203
funding. *See* financial aspects
futuristic thinking. *See* prophecy and futuristic thinking

Galileo, 127–28
Galvin, Bob, 103
Gandelot, Howard, 188
Gates, Bill, 92, 210
Gehan, Tully, 220–21
gender bias, 162–63
gene chip, 70–71
General Motors (GM), 170, 188
genetic disorders, 65–70
genomics, 66, 93
Gernsback, Hugo, 137
Gershenfeld, Neil, 206–7
Gertner, Jon, 170
Geschke, Charles, 117
Getzels, Jacob, 56–58
Ghana, 27
Gibson, William, 216–17
glue gun, 31, 175
Go-Betweens, 157–67
Google, 19, 175, 177
Gould, Chester, 104
grain mill, motorized, 27

Gratzer, Walter, 49
Grayson, Tim, 85–86
Griffin, Abbie, 29, 171
Griffith, Saul, 207–12
groupthink, 150
Guilford, J. P., 149
Guyana, 207–8

Haber, Fritz, 92, 238n
Haber-Bosch process, 92, 93
hackerspaces (open labs), 59–60, 87–92, 175, 205, 206, 214–15
Haiti, 23, 24, 25–26, 27, 28
Harman, Willis, 132
Harrison, John, 158, 159, 160
Hawker, Chris, 40–42, 221–22
helicopter, 110
Herbert, Lawrence, 80–82
Hodak, Max, 75
holistic invention, 180–89
Hollinger, Steve, 51–53
hologram technology, 38, 152–53
How to Learn to Invent (Altshuller), 142
Howtoons, 210–11
Hubbard, Al, 126
Hull, Chuck, 172–74, 176

Ideal Final Result, 147–48
imagination
  ability to imagine failure, 34–35, 36, 61
  to alleviate frustration with repetitive tasks, ix–x, 4–5, 10–11, 32, 117
  concentration to enhance, 127–29, 136, 213
  craftsmanship of the, xi–xii, 124, 135–36, 202
  drugs to enhance, 125–27, 130–35
  as an escape, 201, 203
  experiments using the (*see* thought process, mind experiments)
  Ideal Final Result, 147–48
  mental blocks to, 144, 149–52, 164–65, 193
  the mind's R&D lab, 125–36, 213
  psychology of, 56–58, 61, 84, 116, 118, 124, 131
  role in invention, xvi, 56–58, 61, 80, 105, 110, 123–24, 213
  seeing value in nothing, 80, 82, 84, 92–94 (*see also* recycling and repurposing)
Indiegogo, 36, 39, 40, 222
information theory, 169
InnoCentive, 159–64, 165, 184–85, 186
innovation, vs. invention, xi–xii
inspiration
  source of, xiv–xv, 82, 219–20
  through serendipity, xvi, 47–49, 53–56, 68
Intel Corporation, 138, 215
International Foundation for Advanced Study, 131
Internet
  community creation through, 14, 19, 21, 177, 179, 221
  direct sales through the, 32, 221–22
  financial backing through the (*see* crowdfunding)
  links, precursors to, 110, 114
  as a negative space, 14
  precursor to the, 115, 170
  use in bioinformatics, 62, 69
  use in cancer screening tool, 214–15
  use in drug development, 74–76, 77
  YouTube, 14, 19, 106
  as a zone of permission, 170
intuition, 68, 77
invention
  accidental (serendipity), xvi, 47–49, 53–56, 68
  centralized vs. decentralized, xiv, 21, 73, 168–69, 175–79, 219–20, 222–23
  elegance in simplicity, 146

of experience, vs. a product, 98, 107, 114, 117–18
holistic, 180–89
vs. innovation, xi–xii
merging of diverse disciplines, xvii, 158–61, 164–65, 166, 172, 185–86, 219–20
necessity is the mother of
to escape repetitive tasks, ix–x, 4–5, 10–11, 32
to escape torture, 196–98
limitations of adage, xvi
process (*see* research on the invention process)
reinvention of the process, 222–23
science of inventing (inventology), 138–49, 202
use of term, xi, 30
inventors
consultant company for, 40
demographics, 176
fieldwork by, 23–30, 85–86, 101, 172, 178–79, 208
general public, 14–16, 21
historical background, 168–69
hobbyists, 14, 118
percent in general public, 15
percent of patent holders with day-job discovery, 49
perseverance as most common characteristic, 193
Serial Innovators, 29–30, 171–72
spatial-thinking ability of, 213
unethical, 42
iPhone, 177, 183
ISEF Gordon E. Moore Award, 215

Jarvis, Jim, 86
job, advantages of losing, 165, 166–67
Jobs, Steve, 118
Joel, Amos, 102
Johansson, Patrik, 145
Johnson, Brian David, 138
Johnson, Lonnie, 47–48

Kahn, Herman, 121, 241n
Kahneman, Daniel, 150–51
Kay, Alan, 115, 116, 123
Kekulé, August, xv
Kelly, Kevin, 122
Kennedy, Pagan, 129, 135–36
Kickstarter, 36, 37–38, 39–40, 221, 232n
Kitty Litter Machine, 89–90
Klein, Calvin, 81
Klein, Gary, 34–35, 36
Klein, Marty, 59
Krolopp, Rudy, 103
Kust, Sue, x

labs
Bell Labs, 101, 102, 106, 169–71
the mind's R&D lab, 125–36, 213
MIT Media Lab, 206, 208
open (hackerspaces), 59–60, 87–92, 175, 205, 206, 214–15
Lakhani, Karim, 159–60, 162–63, 177
Lanier, Jaron, 121
Larami, 47
laser technology, 61–62, 133, 210
Latour, Debra, 16–18
Laughlin, Tom, 165–67
Lead Users, ix–x, 4–8, 41, 99, 118, 180–87, 223, 229n
Leboeux, 142
Lemelson-MIT National Collegiate Student Prize, 27, 209
Librium, 132
Lifshitz-Assaf, Hila, 178–79
lights
kayak light, 53
laser, 61–62, 133, 169
strobe, 59
ultraviolet, 172
Lindberg, John, 50
longitude, calculation of, 158, 159, 160

Looking Glass, 38
LSD, 126–27, 130–34, 243n
Lubinski, David, 213
luck
  the art of, 56–58
  being highly observant as, 54
  being open to serendipity as, 55–56
  as a form of creativity, 51–53
luggage, rolling, 3–4, 9

MacGyvers, 85–86, 185–86
Magenis, Ellen, 67–68
magnetic tape, 126, 134, 171
Makri, Stephann, 55–56
Malyshev, Capt., 196, 197–99
Mars, 3–4, 9, 99
Marx, Karl, 207
Massachusetts Institute of Technology (MIT)
  Building 20, 87–92
  and Bush, 108–9
  and Frayne, 28
  and Gershenfeld, 206–7
  and Griffith, 209–11
  Lemelson-MIT National Collegiate Student Prize, 27, 209
  Media Lab, 206, 208
  Sloan School of Management, 5
  and Smith, 23–28
  Strobe Alley, 59–60
Maxwell, Scott, 7–8
Mazur, Robert H., 48
medicine
  biotechnology, 61, 74–76, 183–85, 220
  birth-control pill, 93
  cancer research, 72, 73, 76–77, 214–15
  catheter, Fogarty, 185–86
  catheter-infection biosensor, 181–87
  dentistry, cosmetic, 164
  diabetes, 163–64
  discovery as a byproduct of other research, xvi, 63
  drug development
    big pharma vs. little pharma, 73–78
    costs, 62, 65, 77
    data mining for, 64–65, 71–73, 77
    from plants, 93
    targeted searches, 63–64
    traditional trial-and-error hypothesis-testing, 77
  eyeglass lens fabricator, 207–9
  genetic data, 65–73
  genomics, 66
  gut infection, 93
  InnoCentive challenges, 161–62, 184–85, 186
  laser surgery, 62
  patient safety in hospitals, 180
  penicillin, xvi, 80
  PortaTherm incubator, 25
  predictions, 122, 137–38
  preeclampsia, early warning of, 72
  prosthetic devices, 16–20, 230n
  Smith-Magenis Syndrome, 66–70
  surgical tools, 5, 15
  talismans, 217
  tricorder of *Star Trek*, 137–38
  user-inventors, 5, 15, 180–87
  Viagra, 63
mescaline, 131, 243n
microfilm, 108–9
microtome, 134
mind experiments. *See* thought process, mind experiments
miniaturization of technology, xv, 103, 113, 118, 119–21
mirror, fogless, 31, 32
modification. *See* tinkering
Molquant, 70
Moore, Gordon, 119, 120–21, 122–23, 146
Moore's Law, 120–21, 122–23, 130, 146
Moravec, Christine, 183–84, 186
Mother of All Demos, 115
Motorola, xv, 98–105, 107, 144
Mount Sinai Hospital, 98–100
movies. *See* films

Mumford, Michael, 152–53
Musemeche, Catherine, 182
music, Rat distortion pedal, 60
Musk, Elon, 129

Nader, Ralph, 187–88
nanotechnology, 147, 211–12, 215
NASA, 111, 113, 159, 178–79
National Science Foundation (NSF), 14, 214
navigation, in the open sea, 157–58
negative spaces, 13–14
Nelson, Reed E., 85
Nersessian, Nancy, 60–61
Ness, Roberta, 218
Netflix, 71
Netherlands, xii
NuMedii, 74

observation
  discovery of Smith-Magenis Syndrome, 67, 68
  "luck" as being highly observant, 54
  role in invention, xi, xv, 30, 68
Odeo, 9
open labs, 59–60, 87–92, 175, 205, 206, 214–15
open systems, 158, 161, 204, 219–23
Osterloh, Ian, 63
Owen, Ivan, 19

pager, 98–100
Pantone Matching System, 80–82
particle accelerator, 134
patents
  city with most per capita, xii
  copyright issues, 6, 17, 19
  data mining of, xii, 139–40, 195, 204
  for imaginary technologies, 142
  most-patented inventor, 91–92
  unsought, as negative spaces, 13–14
Paxton, Bill, 115
Pearsall, Duane, 49–51
perceptions of others. *See* feedback
perseverance, 193, 203
petabyte, 71, 77
Pfizer, 63
phones. *See* cell phones
phone systems. *See* AT&T, Bell Labs
photography, strobe, 59
Picasso, 165
Pixar, 123
Plath, Robert, 3–4
play, 51, 57, 59, 89, 91, 169
PlayStation, 106
Playtex, 24, 33
Pong Effect, 97–107
Pong (video game), 106–7
Popular Mechanics (magazine), 29, 210
Post-it Note, xi
pre-mortem (ability to imagine failure)
  crowdfunding as, 36–42
  process, 34–35, 61
Price, Raymond, 29
printers
  to generate solar panels, 37
  hologram technology, 38
  ink for, color standardization, 80–82
  3-D, 19–20, 88–92, 141, 173–75, 214, 221–22, 248n
problem finding
  feedback, the future of, 31–43
  overview, xvi
  through expertise, ix–x, 4–5, 10–11
  through fieldwork, 23–30, 85–86, 100–101, 172, 178–79, 208
  by user-inventors, 12–22
Pro Co Sound, 60
product development. *See* research and development
prophecy and futuristic thinking
  creation of an experience, vs. a product, 98, 107, 114, 117–18
  Delphi priestess, 127
  fictional works (*see* science fiction, as predictive)

prophecy and futuristic thinking (*cont.*)
imagining failure to then adapt product, 34–35, 61
by Lead Users due to expertise, 7–8, 9, 41, 99, 118, 180–87, 223
the mind's R&D lab, 125–36, 213
miniaturization of technology, xv, 103, 113, 118, 119–21
Moore's Law, 120–21, 122–23, 130, 146
overview, xvi–xvii, 97–98, 107, 111, 136
Pong Effect, 97–107
reliability of, 121–24
scalability, 119, 120, 122
time-travel, 137–53, 216–17
by un-thinking an existing condition, 82, 116, 150–52
use of extrapolation, 110–11, 112, 118, 123, 146
Wayne Gretzky Game, 108–24
prosumer, use of term, 21
prototype
to fabricate parts, 173–74
to market product, 32, 36, 209, 220–22
"science fiction prototyping," 138
to test function of invention, xi, 166, 185, 220–21
psychology
aversion to the inner world, 135
being open to serendipity, 55–56, 68
don't worry about perfection, 89
empathy, 23–30
frustration with repetitive tasks, ix–x, 4–5, 10–11, 32, 100, 117
human memory transformation, 109, 110
humans shaped by the invention (Pong Effect), 97–107
hypergraphia, 82
of imagination, 56–58, 61, 84, 116, 118, 124, 136
inertia, 145
intuition, 68, 77
learning from feedback, 34–35
mental blocks, 144, 149–52, 164–65, 193
mindset of continuous improvement, 34
mindset of problem-solving, 51, 86, 91, 116, 145–46, 153, 202
over-investment, 35
self-criticism, 193
sure recognition when it is right, 41, 48
un-thinking an existing condition, 82, 116, 150–52

R&D 100 Awards, 175–76
Raskin, Andy, 144
Rat distortion pedal, 60
recycling and repurposing, 83, 85–92, 166, 208–9
refrigerator, evaporation-based, 28
research and development (R&D)
collaboration
advantages of, 158, 218, 219–20
through crowdfunding, 36–42
in zones of permission, xi, 169–70, 177
contracted lab services/equipment, 74–75
design process, 41, 103, 149–51, 187–89
drug (*see* drug development)
feedback through crowdfunding response, 37–42
finding a need for preexisting solution, 48–53, 61–62, 164
future scenarios, xiii–xiv, 176–78, 222–23
hackerspaces and open labs, 59–60, 87–92, 175, 205, 206, 214–15
lab, in the mind, 125–36, 213
for mass-marketing and distribution, xi, 6–7, 42, 209, 221–22

prototype (*see* prototype)
through the Internet community, xiii, 14, 19, 21, 177, 179, 221
research on the invention process
analysis of R&D 100 Awards, 175–76
historical, xii–xiii
imagination, 56–58, 60–61, 149
interviews of Serial Innovators, 29–30, 171–72
inventology by Altshuller, 138–49, 202
mental blocks, 144, 149–52, 164–65, 193
observation, 54
reliability of predictions, 121–22
repurposing of existing items/conditions, 85–86
through problem-solving competitions (*see* competitions)
by von Hippel, 6, 7, 229n
reverse engineering, 111
Rivers, Adam, 160–61, 165
Robinson, Kelly, 65–67, 68–69
Robinson, Murray, 65–67, 68–70
Robohand, 19
robotics, 7, 8, 19, 61, 75, 77, 88, 180
Roddenberry, Gene, 104–5, 138
Roscoe, Jim, 86
Rossman, Joseph, 193
Russell, James, 133

Saarinen, Eero, 169, 170
Sachs, Ely, 88–89
Sadow, Bernard D., 3, 4, 9
Samsung, 205
satellite transmission, 169
Sauntharajah, Yogen, 76–77, 181–85, 218
scalability, 119, 120, 122
scams, 42
Schimpff, Stephen, 186
Schlatter, James, 48
Schull, Jon, 18–20
science fiction, as predictive
and Altshuller, 141–43, 194, 201
Dick Tracy gadgets, 99, 104
hand-held phone, xv–xvi, 104–5
The Jetsons phone design, 103
and Jules Verne, xvi–xvii, 139, 142, 194
Memex memory machine, 109–11, 112, 153
registry of ideas in, 142
Star Trek, 104–5, 137–38, 206
use of extrapolation, 110–11
William Gibson's worldview, 216–17
scientific research
basic, 60–62, 63, 220
science of inventing, 138–49, 202
Searle, 48
Seinfeld, Jerry, 79
Senegal, 27
serendipity
anti-serendipity factors, 64
being open to, 55–56, 68
engineered, with data mining, 70–73
role in invention, xvi, 47–49, 53–56, 68
Serial Innovators, 29–30, 171–72
Shapiro, Rafael, 139–40, 148, 195
Short Message Service (SMS) protocol, 9
Shovell, Adm. Cloudesley, 157
Sims, Nat, 6–7, 15
Smith, Adam, 4–5
Smith, Alvy Ray, 123
Smith, Amy, 23–28
Smith, Ann, 67–68
Smith, Steven, 149–51
Smithsonian, 214
smoke detector, 49–51
social activism, 187–89, 201, 203, 208, 218, 223
social injustice, 207, 217, 218, 223

social media, 165. *See also* Facebook; Twitter
Solomon, Kenn, 13
sound recording, 126, 132–33, 134, 171
Soviet Union, xii, 138, 139, 141, 143, 194–204
space exploration
  influence of Jules Verne, xvi–xvii, 93
  International Space Station, 179
  Mars, 7–8
  rocket research, 126
  solar-particle storms, 178–79
spatial reasoning, 213
spiritual aspects, 126
sports
  mountain bikes, 5
  team mascots, 12–13
  tennis ball hopper, ix–x
  tennis ball inflator, 31, 33
  T-shirt cannon, 12–13
Spradlin, Dwayne, 161–62
Stalin, Joseph, 141, 194–95, 201
Stanford Research Institute (SRI), 113–15, 132
Stanford University, 131, 132, 162, 175
Stap, Jake, ix–x
steampunk, 216–17
Stolaroff, Myron, 125–27, 129–34, 204, 243n
Stratasys, 175
submarine, 142
sucralose, 49
suitcase, rolling, 3–4, 9
Super-Encounterers, 47–62, 68, 76
Superhero Cyborg workshop, 17–18, 230n
Super Soaker, 47–48
sweeteners, artificial, 48–49, 90–91

tan, artificial, 165–66
tape, magnetic, 126, 134, 171
technology
  low-cost, for developing countries, 24, 27, 29
  miniaturization of, xv, 103, 113, 118, 119–21
  principles of evolution, 140–42, 146
  speed of evolution and Moore's law, 120–21, 122–23, 130, 146
  in *Star Trek*, 104–5, 137–38, 206
Ted Talk, 215
television
  engineering game show, 204
  The Jetsons, 103
  Seinfeld, 79
  Star Trek, 104–5, 137–38, 206
  technology of, 141–42, 152
Tesla, Nikola, xi–xii, 128–29, 137, 203, 213
thought process
  associative nature of, 110
  contradiction paradox, 145–46, 202
  dreaming, xi, xv, 105, 129
  groupthink, 150
  inventive problem solving (TRIZ), 143–49, 195, 201–5
  law of least effort, 150–51
  meditation or illumination-phase, 17, 133
  mental iteration, xi–xii, 34–35, 61, 127, 144
  merging of diverse ideas/disciplines, xvii, 158–61, 164–65, 166, 172, 185–86, 219–20
  mind experiments
    of Altshuller, 141, 142, 143, 144, 147–48, 201
    ball hopper, ix–x
    of Einstein, 128
    of Galileo, 127–28
    helicopter, 110

historical context, 127–29
holographic television, 152–53
Ideal Final Result, 147–48
Memex, 108–11, 123, 153
spatial-thinking ability, 213
of Tesla, 128, 218
out-of-the-box, 177
scientific research on the (*see* research on the invention process)
strong thinking, 213
See also imagination; inspiration
time travel, 137–53, 216–17
tinkering (data algorithms), 69–70, 71, 72
tinkering (mechanical)
by early settlers, 168
education in creative thinking through, 211–13, 214–15
etymology, 58
MacGyvers, 85–86, 185–86
modification of existing products, 15, 20, 114, 230n
recycling/repurposing, 83, 85–92, 166, 208–9
using failure to then discover product, 59–60, 169
tinkering (mental iteration)
imagining failure to then adapt product, 34–35, 61
learning to use, xi–xii, 124
within an imaginary world (*see* thought process, mind experiments)
toothbrush, 82, 217
Townes, Charles H., 61–62
toys and games
Carbon Flyer, 222
Howtoons for instructions to make, 210–11
PlayStation, 106
Super Soaker, 47–48
video games, 106–7
Toy Story (film), 123
Transcriptic, 75
transistor, xv, 15, 100, 103, 107, 169
transportation
cars, 9, 145, 151–52, 166, 174, 187–88
early train design, 151
helicopter, 110
predictions, 122, 129
Trettel, Ben, 212–13
Trident Design, 40
TRIZ (theory of inventive problem solving), 143–49, 201–5
Tully (child), 20
Twitter, 9–11, 36–37, 175

Uber, 76
Unsafe at Any Speed (Nader), 187–88
U.S. Department of Agriculture, 214
U.S. Department of Education, 214
usefulness/uselessness, 30, 58, 84–85
user-inventors
community of, 11, 12–14, 17–20, 37, 230n
comprising dark matter of invention, 14–22
Lead Users, ix–x, 4–8, 41, 99, 118, 180–87, 223, 229n
negative space of sharing by, 13–14

Van As, Richard, 19
Vavilov, Nikolai Ivanovich, 194
Verne, Jules, xvi–xvii, 93, 139, 142, 194
video cameras
aerial, 51–52
video games, 106–7
video recorders
by Ampex, 125, 126, 130–31
Viridis3D, 91
vision
in computer design, 113, 118
in education of invention, 210–11

vision (*cont.*)
multiple times existing simultaneously, 216–17
role in invention, xvi, 110, 112, 123–24
Vojak, Bruce, 29
von Hippel, Eric, 5–6, 7, 14, 177, 229n

Wallas, Graham, 51
Wall Street (film), 104
watchdogs of design, 187–89
water quality testing, 23, 24–25
Wayne Gretzky Game, 108–24
weapons, 139, 238n
Wenger, Brittany, 73
whistleblowers, 187–89, 223
"Who Made That?" column, ix
Windbelt generator, 29
Wiseman, Richard, 54
women's issues, 162–63
Wu, Tim, 106, 171
Wunderman, Irwin, 131

Xerox and Xerox PARC, 84, 115–18, 123, 144, 170

Yahoo!, 175
yarn recycler, 83–84
Y Combinator, 75
The Year 2000 (Kahn and Wiener), 121
Young, Lewis, 120
YouTube, 14, 19, 106

Zasetskii (cell mate of Altshuller), 197, 198
Z Corporation, 91
zones of permission, 168–79

Pagan Kennedy is a writer, journalist and former Knight Science Journalism Fellow at MIT, where she studied microbiology and neuro-engineering. She also held a Lemelson Fellowship at the Smithsonian Institute and is the author of several works of fiction and non-fiction. Her journalism has appeared in the *New York Times Magazine, Boston Globe Magazine,* the *New York Times Book Review* and other publications.